College Trigonometry

College Trigonometry

Marvin Marcus · Henryk Minc
THE UNIVERSITY OF CALIFORNIA AT SANTA BARBARA

HOUGHTON MIFFLIN COMPANY · BOSTON
NEW YORK · ATLANTA · GENEVA, ILLINOIS · DALLAS · PALO ALTO

Library of Congress Catalog Card Number: 73-143323

ISBN: 0-395-12059-4

To

Avraham and Dina Singler

Preface

This text is intended for a standard course in trigonometry. The book is essentially self-contained, though it is assumed that the student has had the equivalent of a one and a half year course in high school algebra and a year of geometry. The complete book can be covered by an average freshman in one quarter or one semester.

There are about one hundred eighty worked examples in the text. Full solutions to about a third of the exercises in the book form a wealth of additional examples illustrating the material. Each section closes with a set of exercises and a true-false quiz which is aimed at developing the student's reasoning capacities and testing his understanding of the definitions and theorems.

In the first two chapters, which are introductory, we develop the prerequisite material on sets, number systems, functions, graphs, etc. In Chapter 3 the student is introduced to trigonometric functions by means of right-angled triangles and their generalization as functions of general angles. We feel that this approach will seem more natural to the student than the more fashionable "modern" method of first defining the circular functions. We introduce circular functions in Chapter 6 as a parallel concept to trigonometric functions. Chapters 4 and 5 contain standard material on trigonometric identities and solution of triangles. Techniques are explained in detail, numerous examples of varying degrees of difficulty are included, and a number of applied problems are discussed, e.g., elementary applications of trigonometry to vectors. The last chapter contains introductory material on complex numbers and their trigonometric form, including De Moivre's theorem.

The present book covers roughly the same material as Chapters 1, 2, 6, and 7 of our *Algebra and Trigonometry*. The content of this book, however, reflects numerous suggestions we have received since the appearance of that text: the material has been substantially expanded, the exposition in many places has been improved and simplified, and a large number of additional worked examples and exercises have been added.

The authors would like to express their appreciation to Miss Barbara Smith for her invaluable assistance in the preparation of this book. Mrs. Sonia Ospina did an admirable job of typing the manuscript.

Marvin Marcus

Henryk Minc

Contents

Sets and Numbers

1.1 Sets and Subsets

A *set* is one of the most elementary and primitive concepts in mathematics. The word "set" will not be formally defined in terms of more elementary ideas. However, some synonyms for the word are collection, aggregate, totality, class. Consider the following examples:

(a) the set consisting of the numbers 0, 1, 2;

(b) the set of all blue-eyed coeds at the University of California at Santa Barbara;

(c) the set of all letters appearing in the word "banana";

(d) the set of all even numbers between 2 and 10, including 2 and 10;

(e) the set of all grains of sand on Goleta Beach;

(f) the set of all electrons in the universe;

(g) the set of all whole numbers.

In order to discuss these examples, we introduce notation that will be fundamental throughout this book. First is the notation for set membership. If S is a set and x is an item that belongs to this set, we write

$$(1) \qquad\qquad x \in S.$$

The formula (1) is read "x is a member of S," or "x belongs to S," or "x is an *element* of S." It means, then, that x is one of the items in the set S. It is clear that we can fully describe a particular set if we are able to write down explicitly each one of the elements (members) in S. In example (a) above, where S is the set consisting of the numbers 0, 1, 2, we write

$$0 \in S, 1 \in S, 2 \in S.$$

1

In example (b), we can list all the elements of the set if we can write down the names of all blue-eyed coeds at U.C.S.B. If the set of letters in the word "banana" is denoted by S, then

(2) $b \in S, a \in S, n \in S.$

Of course, a and n both appear more than once in the word "banana," but we are interested only in the distinct letters in the word, and unless we wish to specify also the number of times each of these letters occurs, we would not write them down more than once. The set S in (d) is completely specified by

(3) $2 \in S, 4 \in S, 6 \in S, 8 \in S, 10 \in S.$

The sets described in (e) and (f) defy any kind of explicit enumeration of the type that we used in (3). In principle we could label each one of the grains of sand on Goleta Beach in some way and thereby describe the set. In the case of all electrons in the universe, however, such a labeling process is completely out of the question, even though we can still believe that this set is a comprehensible thing.

In example (g) there is no way, even in principle, in which we could write down all the whole numbers explicitly. It is true, though, that given any object, most of us can decide whether or not it is a whole number and therefore an element of the set of whole numbers.

We see from these examples that there are essentially two ways to describe a set. First, we can explicitly name all the elements of the set, or second, we can describe the set in terms of some common defining property of all the elements. These two methods of defining sets lead us to introduce some standard useful notation. Curly brackets will be used to denote sets in the first instance. For example, in (a) we can write

(4) $S = \{0, 1, 2\}.$

The set in (c) above can be written

(5) $S = \{b, a, n\}.$

The set in (d) can be written

(6) $S = \{2, 4, 6, 8, 10\}.$

No order is implied in this notation. Thus (6) could also have been written

$$S = \{4, 6, 8, 10, 2\},$$

and (5) could have been written

$$S = \{a, b, n\}.$$

The second way of describing a set is as follows:

(7) $S = \{x \mid x$ satisfies the defining property$\}$.

The formula in (7) is read "S is equal to the set of all x such that x satisfies the defining property." Thus the set S in (a) could also have been denoted

$$S = \{x \mid x = 0 \quad \text{or} \quad x = 1 \quad \text{or} \quad x = 2\}.$$

We could write the set in (d) as follows:

(8) $S = \{x \mid x = 2k, k = 1, 2, 3, 4, 5\}.$

The formula (8) is read, "S is equal to the set of all x which satisfy $x = 2k$, where k is any one of the numbers 1, 2, 3, 4, or 5."

There is nothing mutually exclusive about the two methods of describing a set. We can use either one depending on circumstances; but as we have seen, it may not always be possible to describe a set by the first method.

We shall also find it convenient to have a notation which tells us when an element is *not* a member of a given set. For instance, $\frac{1}{2}$ is not a member of the set of all whole numbers, and we can indicate this notationally by putting an inclined stroke through the set membership symbol,

(9) $\frac{1}{2} \notin S.$

The formula (9) is therefore read "$\frac{1}{2}$ is not an element of S."

We say that two sets are equal if they are the same set. Thus S and T are equal if S and T consist of precisely the same objects. Of course, a set may be described in various ways, and it is not always trivial to assert the equality of two sets. For example, consider the following two descriptions of the same set:

$$S = \{x \mid x \text{ is a whole number between 1 and 50 and } x \text{ is not}$$
$$\text{divisible by any whole number except 1 and } x\}.$$

It takes a moment's reflection to see that

$$S = \{1, 2, 3, 5, 7, 11, 13, 17, 19, 23, 29, 31, 37, 41, 43, 47\}.$$

The set T consisting of the integers 2, 8, 10 is part of the set S in (6). We denote this situation notationally as follows:

(10) $T \subset S.$

The formula (10) is read "T is contained in S," or "T is a *subset* of S," and the above symbol is called an *inclusion* sign. It means simply that each member of the set T is a member of the set S; equivalently, for each x, if $x \in T$, then $x \in S$. There is an important distinction in meaning between the set membership sign $\in$ and the inclusion symbol $\subset$. Certainly we could not write $T \in S$, for T itself is not any one of the numbers 2, 4, 6, 8, 10. Nor could we write $2 \subset S$, for 2 is not a subset of S. This second distinction is somewhat more subtle, for we have distinguished between the elements of a set and the set consisting of those elements. Thus it is perfectly clear and indeed correct to write

(11) $\{2\} \subset S,$

for, each member of the set on the left of the inclusion sign is a member of the set on the right of the inclusion sign in (11). Of course, the set on the left in (11) has only one member, namely 2.

As stated above, two sets S and T are *equal* if and only if they consist of precisely the same objects. Equivalently, to say that S is equal to T means that every element in S is in T and every element in T is in S: $S \subset T$ and $T \subset S$.

Consider the following phrase which purports to describe a set: "The set of all whole numbers no larger than 10, but greater than 15." While it makes sense to talk about the set of all whole numbers less than 10 and the set of all numbers greater than 15, there are no numbers which belong to both of these sets. Nevertheless, in mathematics it is convenient to have a symbol which stands for the set which has no members, called the *empty set*. This set often arises in combining perfectly well-defined sets. We shall denote the empty set by

(12) $\phi.$

It is always a true statement that whatever x may be,

$$x \notin \phi.$$

It is also the case that we may always write

$$\phi \subset S$$

(which includes the possibility that S is ϕ!). The above inclusion is always valid, because it is impossible to exhibit an element in ϕ which is not in S. In fact, it is impossible to exhibit an element in ϕ at all!

SYNOPSIS

There are two methods of describing a set: listing all the items and describing the set in terms of a defining property. The notations appropriate for these two methods are

$$\{\ldots,\ldots,\ldots,\ldots\} \quad \text{and} \quad \{x \mid x \ satisfies \ldots\}.$$

The set membership symbol indicates that an item x belongs to a set S. This is written

$$x \in S.$$

If x is not in S, we write

$$x \notin S.$$

The equality of two sets,

$$S = T,$$

means that S and T consist of the same elements. The inclusion symbol $\subset$ as in

$$T \subset S$$

means that each element in T is also in S. The empty set ϕ is the set which has no elements.

QUIZ

Answer *true* or *false:*
1. If E is the set of even whole numbers, then $2\frac{1}{2} \in E$.
2. If M is the set of all whole numbers and E is the set of all even whole numbers, then $E \subset M$.
3. If A and B are subsets of a set S, then the set of all items which belong to either A or B (possibly to both A and B) is a subset of S.
4. If A and B are subsets of a set S, then the set of all items which belong to both A and B is a subset of S.
5. If X is any set, then $X \subset X$.
6. If X is any set, then $X \subset \phi$.
7. If X is any set, then $\phi \subset X$.

8. It is never possible that $\phi \in X$ for any set X.
9. The set which consists of the empty set, i.e., $\{\phi\}$, has no elements in it.
10. $\phi \in \{\phi\}$.

EXERCISES

1. List all subsets of the following set: $S = \{0, 1, 2\}$.
2. List all the subsets of the set of letters appearing in the word "banana."
3. Show by example that an element of a set can also be a subset of the set.
4. Let N be the set of all *natural numbers*, 1, 2, 3, ... Let

$$S = \{x \mid x = n^2, n \in N\}.$$

That is, S is the set of numbers which are squares of whole numbers. Recall that n^2 is just the product of n with itself. For each of the following numbers indicate whether $x \in S$ or $x \notin S$:

$$x = 4, x = 0, x = \tfrac{1}{4},$$
$$x = 10, x = 25, x = 1,000,$$
$$x = 1,000,000.$$

5. Let M be the set of *nonnegative* whole numbers 0, 1, 2, 3, ..., and let

$$E = \{x \mid x = 2m, m \in M\}, \qquad T = \{x \mid x = 3m, m \in M\},$$
$$S = \{x \mid x = m^2, m \in M\}, \quad \text{and} \quad R = \{x \mid x = s + 1, s \in S\}.$$

Insert in each blank space of the following table either YES or NO according as the number on the left is or is not an element of the set above the blank space.

$\in$	M	E	T	S	R	ϕ
9						
5						
0						
6						
26						
-1						

6. Let M, E, T, S, R be the sets defined in Exercise 5. Insert in each blank space in the following table either YES or NO according as the set indicated on the left of the blank space is or is not a subset of the set above it.

$\subset$	M	E	T	S	R	ϕ
M						
E						
T						
S						
R						
ϕ						

7. How many subsets are there of a set which consists of five elements? Can you answer this question without listing all the subsets?

8. Let M be the set of nonnegative whole numbers, $0, 1, 2, \ldots$. Let $S = \{x \mid x = 2m + 3n, m \in M, n \in M\}$; that is, S is the set of all whole numbers which can be written in the form $2m + 3n$, where m and n are nonnegative whole numbers. If E is the set of even nonnegative whole numbers and O is the set of odd nonnegative whole numbers, show that $E \subset S$ but that O is not a subset of S. Show that $S \subset M$. Is it true that $S = M$? If not, exhibit an element of M which is not in S.

1.2 Combining Sets

There are several fundamental operations that can be performed on subsets of a set S to produce other subsets of S. The simplest of these operations is *complementation*. Thus, if S is a set and $X \subset S$, we denote by

(1) $$X'$$

the set of all elements in S which are *not* in X. The subset X' is called the *complement* of X with respect to S. The set S is assumed to be known even though it is not specified in our notation. We can write down the definition of X' as follows:

$$X' = \{x \mid x \in S, x \notin X\}.$$

In other words, X' consists of those elements in S not in X. For example, if S is the set of all nonnegative whole numbers, i.e., S consists of the numbers 0, 1, 2, 3, ..., and if E is the set of even nonnegative whole numbers and O denotes the set of all odd nonnegative whole numbers, then

$$E' = O$$

and

$$O' = E.$$

Also observe that

$$(E')' = E.$$

For $(E')'$, which will henceforth be written E'', is the complement of $E' = O$, and therefore consists of all those whole numbers which are not in O, i.e., $E'' = E$. It is true, of course, that whatever the subset X of a set S may be,

(2) $$X'' = X.$$

To verify (2), just remember that X'' is the complement of the set of objects not in X. In other words, $x \in X''$ means that it is not true that $x \notin X$, i.e., it is true that $x \in X$. We can visualize this situation by means of a simple picture known as a *Venn diagram* (Figure 1.1).

(3)

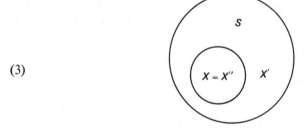

Figure 1.1

Another elementary method for producing new subsets of a set S is the formation of the *intersection* of two subsets. If X and Y are subsets of S, then the intersection of X and Y, denoted by

(4) $$X \cap Y,$$

is the subset of S which consists of precisely those elements which belong to both X and Y. In our set notation we can write

(5) $$X \cap Y = \{z \mid z \in X \ \text{and} \ z \in Y\}.$$

For example, if S is the set of nonnegative whole numbers, $X = \{0, 1, 2, 3\}$, $Y = \{2, 3, 4, 5\}$, then $X \cap Y = \{2, 3\}$. As another example using S, if we take X to be E, the set of even nonnegative whole numbers, and Y to be O, the set of odd nonnegative whole numbers, it is clearly true that $E \cap O = \phi$, since no whole number is both even and odd. This is a special case of the following general statement: If $X \subset S$, then

(6) $$X \cap X' = \phi.$$

We next discuss the operation of forming the *union* of two subsets X and Y of a set S. The union of X and Y, written

$$X \cup Y,$$

is the set of all items which belong to X *or* Y. We use the word "or" in the non-exclusive sense, i.e., the elements of $X \cap Y$ are also in $X \cup Y$. As an example, if S, E, O are the sets we previously discussed, then

$$S = E \cup O.$$

Once again, if $X = \{1, 2, 3, 4, 5\}$ and $Y = \{1, 2, 7, 8\}$, then

$$X \cup Y = \{1, 2, 3, 4, 5, 7, 8\}.$$

It is convenient and informative to construct Venn diagrams to describe unions and intersections of sets. Thus if S is a set and X and Y are subsets of S, consider Figure 1.2. The area included in the shaded regions represents the set $X \cup Y$.

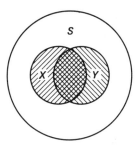

Figure 1.2

The crosshatched area represents the set $X \cap Y$. There are a number of elementary properties of union and intersection that are obvious both from the definitions and from the corresponding Venn diagrams. Let S be a set, and let X, Y, Z be subsets of S. Then

(7) $$\phi \subset (X \cap Y) \subset X \subset (X \cup Y);$$

(8) $$\phi \subset (X \cap Y) \subset Y \subset (X \cup Y);$$

(9) $$X \cap Y = Y \cap X;$$

(10) $$X \cup Y = Y \cup X;$$

(11) $$X \cap (Y \cup Z) = (X \cap Y) \cup (X \cap Z);$$

(12) $$X \cup (Y \cap Z) = (X \cup Y) \cap (X \cup Z).$$

To illustrate the methods of proof, let us verify (11), first by means of a Venn diagram (Figure 1.3) and then directly from the definition.

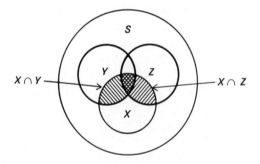

Figure 1.3

The set with the heavy outline is $Y \cup Z$. The lined area is clearly the region $(X \cap Y) \cup (X \cap Z)$.   But this is just precisely the area in which $Y \cup Z$ intersects X.   To give a more formal argument, suppose an element t belongs to the set on the left side of (11), i.e., $t \in X \cap (Y \cup Z)$.   This means that $t \in X$ and $t \in Y \cup Z$, which in turn implies that

$$t \in X \quad \text{and} \quad (t \in Y \quad \text{or} \quad t \in Z).$$

In other words,

$$(t \in X \quad \text{and} \quad t \in Y) \quad \text{or} \quad (t \in X \quad \text{and} \quad t \in Z).$$

Thus

$$t \in X \cap Y \quad \text{or} \quad t \in X \cap Z.$$

Finally,
$$t \in (X \cap Y) \cup (X \cap Z).$$

Hence, if t is any element in the set on the left side of (11), t is in the set on the right side of (11), and therefore

$$X \cap (Y \cup Z) \subset (X \cap Y) \cup (X \cap Z).$$

To obtain equality, it remains to prove that

$$(X \cap Y) \cup (X \cap Z) \subset X \cap (Y \cup Z).$$

Suppose now that t is an element of the right side of (11). Then

$$t \in X \cap Y \quad \text{or} \quad t \in X \cap Z.$$

Hence
$$(t \in X \quad \text{and} \quad t \in Y) \quad \text{or} \quad (t \in X \quad \text{and} \quad t \in Z).$$

In other words,
$$t \in X \quad \text{and} \quad (t \in Y \quad \text{or} \quad t \in Z),$$

i.e.,
$$t \in X \quad \text{and} \quad t \in Y \cup Z.$$

This implies that
$$t \in X \cap (Y \cup Z).$$

Therefore
$$(X \cap Y) \cup (X \cap Z) \subset X \cap (Y \cup Z).$$

If X is a set which has only finitely many elements, let us denote by

$$\nu(X)$$

the number of elements in X. If Y is another set with only finitely many elements and if we want to count the number of elements in $X \cup Y$, i.e., the value of $\nu(X \cup Y)$, then we tabulate separately $\nu(X)$, the number of elements in X, and $\nu(Y)$, the number of elements in Y, and subtract from $\nu(X) + \nu(Y)$, the number of elements included in both sets. In other words,

(13) $$\nu(X \cup Y) = \nu(X) + \nu(Y) - \nu(X \cap Y).$$

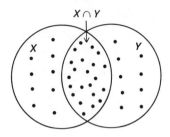

Figure 1.4

We can illustrate (13) with a Venn diagram (Figure 1.4).

To count the number of elements (represented by dots) in $X \cup Y$, we separately count $v(X)$, the number of elements in X, and $v(Y)$, the number of elements in Y. But we have counted the number of elements in $X \cap Y$ twice. Hence we must subtract the last term in (13).

Suppose, for instance, that in a sample of 100 smokers of brands X and Y, 25 people smoke only brand X, and 10 people smoke only brand Y. We can use a Venn diagram to count the number of people who smoke both brand X and brand Y.

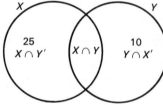

Figure 1.5

We are given that $v(X \cap Y') = 25$ and $v(Y \cap X') = 10$, and we know that $v(X \cup Y) = 100$. It is clear from Figure 1.5 that

$$v(X \cup Y) = v(X \cap Y') + v(X \cap Y) + v(Y \cap X').$$

Hence

$$100 = 25 + 10 + v(X \cap Y),$$

or

$$v(X \cap Y) = 65,$$

i.e., the number of people who smoke both brand X and brand Y is 65.

Example 2.1 A set S consists of 80 integers (whole numbers). If exactly 43 of these are even, 35 are divisible by 3, and 19 are divisible by 6, find the number of odd integers in S that are not divisible by 3.

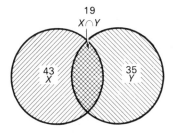

Figure 1.6

Let X be the subset of even integers in S and let Y denote the subset of integers in S divisible by 3. The required number is $v(S) - v(X \cup Y)$. We know that

$$v(X \cup Y) = v(X) + v(Y) - v(X \cap Y).$$

Also, $X \cap Y$ is the set of integers in S that are even and divisible by 3. In other words, $X \cap Y$ consists of all integers in S divisible by 6. Hence

$$v(X \cup Y) = 43 + 35 - 19$$
$$= 59.$$

Thus the number of odd integers in S that are not divisible by 3 is

$$v(S) - v(X \cup Y) = 80 - 59$$
$$= 21.$$

SYNOPSIS

In this section we defined the complement X', intersection $X \cap Y$, and union $X \cup Y$ for subsets of a set S, and we derived various relations among these set operations. The Venn diagram serves as a pictorial device to represent these operations. Finally, we introduced the notation $v(X)$ to denote the number of elements in a set, and noted an elementary fact in (13) about the number of elements in the set $X \cup Y$.

QUIZ

Answer *true* or *false:*

(All sets are subsets of some given set S.)

1. $X \cap X' = \phi$.
2. $X \cup X' = S$.
3. $\phi' = S$.

4. $S' = \phi$.
5. $X'' = S$.
6. $X \cap X = X$.
7. $X \cap Y \subset X \cup Y$.
8. $X \cup X = X$.
9. $X \cup (Y \cup Z) = (X \cup Y) \cup (X \cup Z)$.
10. $X \cap (Y \cap Z) = (X \cap Y) \cap (X \cap Z)$.

EXERCISES

1. Let $X = \{1, 3, 5, 7, 9\}$, $Y = \{1, 4, 9\}$, and $Z = \{3, 6, 9\}$. Represent each of the following sets by listing all elements:
    (a) $X \cup Y$,   (b) $Y \cup Z$,
    (c) $X \cap Z$,   (d) $Y \cap Z$,
    (e) $(X \cup Y) \cap Z$,   (f) $X \cup (Y \cap Z)$,
    (g) $(X \cup Y) \cup Z$,   (h) $X \cup (Y \cup Z)$,
    (i) $X \cap (Y \cap Z)$,   (j) $(X \cap Y) \cap Z$,
    (k) $(X \cap Y) \cup Z$,   (l) $X \cap (Y \cup Z)$.

2. Let M be the set of nonnegative whole numbers, $E = \{x \mid x = 2m, m \in M\}$, $T = \{x \mid x = 3m, m \in M\}$, $S = \{x \mid x = m^2, m \in M\}$, $R = \{x \mid x = s + 1, s \in S\}$. Describe the following sets:
    (a) $E \cup M$,   (b) $E \cap M$,
    (c) $E \cap T$,   (d) $(E \cap T) \cup M$,
    (e) $E \cap (T \cup M)$,   (f) $S \cap R$,
    (g) $S \cap (R \cap M)$,   (h) $S \cap \{x \mid x \in M, x \text{ is less than } 10\}$.

3. Show that

    (14) $$(X \cup Y) \cup Z = X \cup (Y \cup Z)$$

    and

    (15) $$(X \cap Y) \cap Z = X \cap (Y \cap Z),$$

    for any sets X, Y, and Z. Illustrate (14) and (15) by means of Venn diagrams. [*Note:* In view of (14) we can simplify our notation and write $X \cup Y \cup Z$ instead of $(X \cup Y) \cup Z$ or $X \cup (Y \cup Z)$. Similarly, $X \cap Y \cap Z$ will denote either $(X \cap Y) \cap Z$ or $X \cap (Y \cap Z)$. The relations (14) and (15) are called the *associative laws* for union and intersection.]

4. Draw a Venn diagram illustrating the fact that $X \cup X' = S$.
5. Show that $X \subset Y$ is equivalent to the following statement: $X = X \cap Y$.
6. Illustrate (12) by means of a Venn diagram.

7. Illustrate by means of a Venn diagram the following formula: $(X \cup Y)' = X' \cap Y'$. This formula is known as *De Morgan's Law*.

8. Prove the formula in the preceding exercise directly from the definitions.

9. Let X be the set of all whole numbers divisible by 2, let Y be the set of all whole numbers divisible by 3, and let Z be the set of all whole numbers divisible by 6. Show that $Z = X \cap Y$.

10. If X is the set of all whole numbers divisible by 2, Y the set of all whole numbers divisible by 4, and Z the set of all whole numbers divisible by 8, is it true that $Z = X \cap Y$? Explain your answer.

11. Compute $v(X)$ for each of the following sets:
    (a) $X = \{x \mid x \text{ is a positive whole number, and } x^2 \text{ is smaller than } 5\}$,
    (b) $X = \{x \mid x \text{ is a whole number larger than 17 and less than 12}\}$,
    (c) $X = \{x \mid x \text{ is a whole number, } x^2 = 16, \text{ and } 4x = 16\}$.

12. Let $X = \{1, 2, 4, 8, 16\}$, $Y = \{1, 4, 9, 16\}$. Compute $v(X \cup Y)$ and $v(X \cap Y)$ and verify that

$$v(X \cup Y) = v(X) + v(Y) - v(X \cap Y).$$

13. Prove the formula

$$(16) \quad v(X \cup Y \cup Z) = v(X) + v(Y) + v(Z) - v(X \cap Y) - v(X \cap Z) \\ - v(Y \cap Z) + v(X \cap Y \cap Z)$$

and illustrate it by means of a Venn diagram.
[*Hint:* Let $W = Y \cup Z$. Then by (13), $v(X \cup Y \cup Z) = v(X \cup W) = v(X) + v(W) - v(X \cap W)$, and $v(W) = v(Y \cup Z) = v(Y) + v(Z) - v(Y \cap Z)$; therefore, $v(X \cap W) = v(X \cap (Y \cup Z)) = v((X \cap Y) \cup (X \cap Z))$, by (11). Hence, $v(X \cap W) = v(X \cap Y) + v(X \cap Z) - v((X \cap Y) \cap (X \cap Z))$ and formula (16) follows easily.]

14. In a certain high school, 83 students are taking French, 67 are taking German, and 121 are taking Spanish. If 11 students are taking both French and German, 15 students are taking both French and Spanish, 9 students are taking both German and Spanish, and the number of students taking all three languages is 3, find the total number of students who are taking foreign languages.
[*Hint:* Let X, Y, and Z be the sets of students learning French, German, and Spanish, respectively. Apply formula (16).]

15. In a baroque ensemble consisting of 14 players, 8 players can play recorders, 7 players can play viols, and 3 players can play the harpsichord. If 4 players can play both recorders and viols, if one of the harpsichordists can play a recorder, and if two of the harpsichordists can play viols, how many members of the ensemble can play all three kinds of instruments?

16. Let S denote the set of all polygons in the plane. In S, let X denote the set of quadrilaterals, Y the set of squares, Z the set of equilateral triangles, and C the set of polygons having the property that a line segment joining any two points in the polygon is entirely contained within the polygon. Describe in words the following sets:

(a) $X \cap Y$,   (b) $X \cup Y$,   (c) $X \cap Z$,

(d) $X \cap C$,   (e) $Y \cup C$.

1.3   Numbers

The concept of number is as old as human civilization. Despite this, controversy still continues among mathematicians and logicians concerning the precise definition of "a number." In this book and at this level in general, it is inappropriate and probably impossible to present a coherent theory for the system of real numbers. Certain subtleties are involved which require a great deal more mathematical machinery than has been developed at this point. Despite these difficulties, we can discuss properties of the real numbers that should be familiar to students.

Let N denote the set of *natural numbers*. These are just the ordinary counting numbers

$$1, 2, 3, \ldots .$$

The natural numbers are used in everyday life as tags or designations for the number of elements in a set. For example, the number 12 alone does not mean anything to us, although we often think of 12 as a dozen items. We may say that 12 denotes the common property of all sets consisting of a dozen items.

We shall not attempt to define an *integer*. We are really concerned with the behavior of the integers under certain algebraic operations. The fundamental operation involving integers is addition. We all know that the addition of two integers always produces a well-defined integer called the *sum* of the two integers. Historically, this fact led to the introduction of the number "zero" and the so-called negative integers. For, given two natural numbers a and b, we cannot always find another natural number x for which

(1) $$a + x = b.$$

For example, take a and b to be 1. Obviously no natural number x satisfies (1). Thus the symbol zero, 0, evolved to provide a solution to equation (1) in the case $a = b$. This alone, however, did not suffice to make equation (1) solvable for all natural numbers a and b; it was still necessary to increase the set of available numbers by introducing the *negative integers*. These allow us to solve equation (1), for example, when $a = 2$ and $b = 1$. The solution $x = -1$ is a negative integer. Formally, we can simply define "-1" to be the number x which has the property that

$$1 + x = 0.$$

Similarly, if m is any *positive* integer (i.e., any natural number), then the *negative* integer $-m$ is defined by the equality

$$m + (-m) = 0.$$

The set consisting of the natural numbers, zero, and all the negative integers is called the *set of integers*, which will be denoted by Z. The rules for addition and multiplication of natural numbers are extended to all integers. We assume that the reader is familiar with the arithmetic governing Z. In particular, recall that if m and n are positive integers, then

$$
\begin{aligned}
-(-m) &= -1(-m) \\
&= m, \\
0m &= 0, \\
m(-n) &= -mn, \\
(-m)n &= -mn, \\
(-m)(-n) &= mn.
\end{aligned}
$$

(As usual, we denote multiplication by juxtaposition whenever convenient.)

Consider the following problem: Given integers a and b, find a number x for which

$$(2) \qquad\qquad ax = b.$$

The equation $0x = b$, $b \neq 0$, will not have a solution since we know that $0x = 0$ for any x. Thus, we shall assume that a is not zero. Equation (2) presents no difficulties for "nice" situations, such as $b = 51$ and $a = 17$. But even with as simple a choice as $b = 1$ and $a = 2$, we cannot find a solution to equation (2) in the set Z. Thus the set of *rational numbers*, i.e., fractions, was invented in order that equation (2) might have a solution whenever a is not zero. We denote the solution to (2) by

$$(3) \qquad\qquad \frac{b}{a},$$

or by b/a. Two fractions a/b and c/d are equal if and only if $ad = bc$. The totality of fractions

$$Q = \{p \mid p = x/y, \, y \neq 0, \, x \in Z, \, y \in Z\}$$

is called the set of rational numbers, and the operations of addition and multiplication are extended to Q. Observe that $Z \subset Q$.

As a reminder to the reader, the formulas governing the addition and multiplication of rational numbers are

$$\frac{x_1}{y_1} + \frac{x_2}{y_2} = \frac{x_1 y_2 + x_2 y_1}{y_1 y_2}$$

and

$$\frac{x_1}{y_1} \times \frac{x_2}{y_2} = \frac{x_1 x_2}{y_1 y_2} .$$

Reducing a fraction to lowest terms simply means cancelling all common divisors in the numerator and denominator. To subtract and divide fractions, the appropriate formulas are

$$\frac{x_1}{y_1} - \frac{x_2}{y_2} = \frac{x_1 y_2 - x_2 y_1}{y_1 y_2}$$

and

$$\frac{x_1}{y_1} \div \frac{x_2}{y_2} = \frac{x_1 y_2}{y_1 x_2} .$$

Thus, for example,

$$\frac{2}{3} - \frac{4}{7} = \frac{2 \cdot 7 - 4 \cdot 3}{21}$$

$$= \frac{14 - 12}{21}$$

$$= \frac{2}{21} ,$$

and 2/21 is in lowest terms, since the numerator and denominator have no common factors except 1. As another example,

$$\left(\frac{7}{8} \left(\frac{2}{3} + \frac{4}{7} \right) \right) \div \frac{9}{10} = \left(\frac{7}{8} \left(\frac{2 \cdot 7 + 4 \cdot 3}{21} \right) \right) \div \frac{9}{10}$$

$$= \left(\frac{7}{8} \left(\frac{14 + 12}{21} \right) \right) \div \frac{9}{10}$$

$$= \left(\frac{7}{8} \left(\frac{26}{21} \right) \right) \div \frac{9}{10}$$

$$= \left(\frac{7 \cdot 26}{8 \cdot 21} \right) \div \frac{9}{10}$$

$$= \frac{26}{8 \cdot 3} \div \frac{9}{10}$$

$$= \frac{13}{4 \cdot 3} \div \frac{9}{10}$$

$$= \frac{13 \cdot 10}{4 \cdot 3 \cdot 9}$$

$$= \frac{13 \cdot 5}{2 \cdot 3 \cdot 9}$$

$$= \frac{65}{54}.$$

The development thus far is still not adequate to handle even a slight modification in an equation of the form (2). For, we can reformulate the equation (2) as follows: Given a natural number b, find a number x such that

$$xx = b,$$

i.e.,

(4) $$x^2 = b.$$

This problem arises in a natural way in elementary geometry if we want to compute the length of the side of an isosceles right-angled triangle whose hypotenuse is 2 units long (Figure 1.7).

Figure 1.7

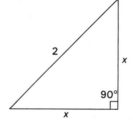

As we know from the Pythagorean Theorem, $x^2 + x^2 = 4$; hence

(5) $$x^2 = 2.$$

Our next result shows that there exists no solution to (5) in Q.

Theorem 3.1 *No rational number x satisfies $x^2 = 2$.*

Proof Suppose x is a rational number satisfying (5), say $x = p/q$. Assume, moreover, that the fraction p/q is in lowest terms (for example, 7/5 rather than 14/10). Now, $x = p/q$ and hence

(6) $$xq = p,$$

where $x^2 = 2$. Since xq and p are the same integer, $(xq)^2 = x^2q^2$ and p^2 are also the same integer. But $x^2 = 2$, and hence

(7) $$2q^2 = p^2.$$

The number $2q^2$ is even, and hence, from (7), the number p^2 is also even. Now, if p were odd, p^2 would be odd (see Quiz Question 4). Thus p must be even. This means that p is twice some integer, say,

(8) $$p = 2m.$$

Hence $p^2 = 4m^2$. If we substitute this in the equation (7), we have

$$2q^2 = 4m^2$$

or

$$q^2 = 2m^2.$$

Thus q^2 is even (it is twice m^2) so that q must be even, i.e.,

(9) $$q = 2n,$$

where n is some integer. Now, (8) and (9) together tell us that the fraction p/q could not have been in lowest terms, because both p and q are multiples of 2. But every fraction can be expressed in lowest terms. The only way out of this dilemma is to admit that our original statement, that $x = q/p$ is a solution of (5), is wrong. In other words, no rational number satisfies (5), and our argument is complete. ∎

In order that (5) have a solution, we are led to invent an even larger class of numbers, the set of real numbers, R. The set R contains all the rational numbers. Real numbers that are not rational are called *irrational numbers*. We have not yet defined a "real number." In order to do this we recall the so-called *decimal representation* of numbers.

Consider, for example, the fraction $\frac{2679}{4747}$. Carrying out the indicated division

```
               .5643
     4747)2679.0000
          23735
          30550
          28482
          20680
          18988
          16920
          14241
           2679
```

we see that our last remainder, 2679, is precisely the number with which we began, and hence the whole division process will repeat itself. Thus

$$\frac{2679}{4747} = .564356435643 \ldots .$$

The recurring group of digits, i.e., 5643, is called the *period* or the *cycle* of the decimal. We write

$$\frac{2679}{4747} = .\overline{5643},$$

where the bar indicates that the group of digits 5643 repeats indefinitely, i.e., that 5643 is the period of the decimal. Similarly

$$4.6\overline{375}$$

denotes the decimal

$$4.6375375375 \ldots$$

whose period is 375. As a matter of fact, a decimal expansion with a period is a characteristic property of a rational number; that is, a number is rational if and only if it has a *repeating* decimal expansion. *Terminating* decimals, being rational numbers, can be interpreted as having a period of length one consisting of the integer 0, for example,

$$.25 = .25\overline{0}.$$

We can easily reconstruct a fraction from a repeating decimal expansion by the process illustrated in the following example. Suppose we want the fraction p/q which is represented by the repeating decimal $1.\overline{612}$. We note that the period of the decimal consists of three digits. If we multiply

$$\overline{.612} = .612612612 \ldots$$

by 1000, we shift the decimal point three places to the right to obtain

$$612.612612 \ldots = 612.\overline{612}.$$

Let

$$r = 1.\overline{612}.$$

Then we have just seen that

$$1000r = 1000 \times (1 + .\overline{612})$$
$$= 1000 + 1000 \times .\overline{612}$$
$$= 1000 + 612.\overline{612}$$
$$= 1612 + .\overline{612}.$$

Thus we have

$$1000r = 1612 + .\overline{612}$$

and

$$r = 1 + .\overline{612}.$$

We can subtract the second equation from the first one to obtain

$$1000r - r = 1612 - 1,$$
$$999r = 1611,$$
$$r = \frac{1611}{999}.$$

The technique exhibited in this example is perfectly general. In fact, this computation in its general form implies that every repeating decimal is a rational number. It follows that an irrational number cannot be equal to a repeating decimal.

We saw in Theorem 3.1 that the square root of 2 is not rational. Thus, by the preceding argument, it cannot have a repeating decimal expansion. We can write down the first few digits in the decimal expansion of $\sqrt{2}$:

(10) $$\sqrt{2} = 1.414213 \ldots .$$

Equation (10) means that $\sqrt{2}$ is greater than 1.414213 and less than 1.414214. We can compute the decimal expansion of $\sqrt{2}$ to any desired degree of accuracy, but the resulting decimal does not repeat. Indeed, it is unlikely that anyone knows the one-billionth digit in this decimal expansion.

Definition 3.1   (Real Number)   A *real number* is a decimal, repeating or nonrepeating. The set of real numbers together with the operations of addition and multiplication will be denoted by *R*.

Defining the real numbers presents some serious difficulties whose resolution requires a very intricate development. This book will not attempt to uncover and investigate the subtleties inherent in the definition of a real number.

The real numbers are separated into three mutually exclusive sets, the set of *positive* real numbers, which we shall denote by *P*, the set of *negative* real numbers, and the set consisting of the number zero. The set *P* possesses two important properties:

(*i*) if $a \in P$ and $b \in P$, then $a + b \in P$;
(*ii*) if $a \in P$ and $b \in P$, then $ab \in P$.

Definition 3.2 (Inequality) We say that *a is greater than b*, written $a > b$, if the difference $a - b$ is positive, i.e., $a - b \in P$. It is sometimes convenient to allow for the possibility that *a* and *b* are equal, i.e., $a - b = 0$; we then write $a \geq b$. These relations are also written $b < a$ or $b \leq a$, which are read *b is less than a* or *b is less than or equal to a*, respectively. If *x* is a real number, then

$$a \leq x \leq b$$

means that $a \leq x$ and $x \leq b$. The relations "greater than" and "less than" are generally referred to as *inequalities*.

The set of real numbers *R* can be represented geometrically on the so-called *real number line*. This representation assigns to each real number *r* a point on a line as follows. We draw a line and designate two distinct points to represent the numbers 0 and 1. Let the point representing 1 be to the right of the point representing 0. The length of the segment from 0 to 1 is used as the unit. We now lay off successive intervals of unit length to the right of 0 and assign to the successive endpoints the integers 1, 2, 3, 4, ... We then lay off successive unit intervals to the left of 0 and assign to the endpoints of these intervals the integers -1, -2, -3, -4, . . . in order. We thus obtain a representation of integers by uniquely assigned points on the line (Figure 1.8).

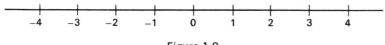

Figure 1.8

In order to represent finite decimals by points on the line, we suitably subdivide the interval between two points representing integers. The process is best described by an example. Consider the representation of the number 1.72 (Figure 1.9).

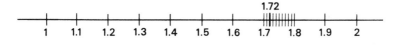

Figure 1.9

The number 1.72 is located by subdividing the interval between 1 and 2 into ten equal parts and letting these stand for 1.1, 1.2, 1.3, etc., then further subdividing the interval between 1.7 and 1.8 into ten equal parts representing 1.71, 1.72, 1.73, etc.

In order to represent a nonterminating decimal *r* we construct a nested set of intervals of decreasing length: Each successive digit in the decimal expansion of *r* defines the subinterval in which the point representing r must be located. Thus, for example, the point representing $\sqrt{2} = 1.414213\ldots$ must lie in the intervals

$$
\begin{aligned}
&\text{between} \quad 1 \qquad\;\; \text{and} \quad 2; \\
&\text{between} \quad 1.4 \qquad \text{and} \quad 1.5; \\
&\text{between} \quad 1.41 \qquad \text{and} \quad 1.42; \\
&\text{between} \quad 1.414 \quad \text{and} \quad 1.415, \text{ etc.}
\end{aligned}
$$

To each such real number r corresponds therefore a nested set of intervals closing down around the point which represents r on the real line. It must be considered an *axiom* fundamental to both the real numbers and the geometry of the line that once the points which represent 0 and 1 are prescribed and a unit length thereby determined, then *to each point on the line there corresponds precisely one real number and to each real number there corresponds precisely one point on the line.*

SYNOPSIS

In this section we introduced the following sets of numbers:

N,   the set of natural numbers;
Z,   the set of integers;
Q,   the set of rational numbers;
R,   the set of real numbers.

The following inclusions hold for these sets:

$$N \subset Z \subset Q \subset R.$$

We established a correspondence between the points on the line and real numbers.

QUIZ

Answer *true* or *false:*
1. -1 is a natural number.
2. $1 + \sqrt{2}$ is a rational number.
3. If k^2 is an odd natural number, then k is an odd natural number.
4. If k^2 is an even natural number, then k is an even natural number.
5. $2.7 \in Q$.
6. $\sqrt{2} \in R$.
7. $\sqrt{3} > \sqrt{2}$.
8. $2.3\overline{8} \le 2.\overline{38}$.
9. Two fractions a/b and c/d are equal if and only if $a = c$ and $b = d$.
10. If $z \in R$, then $z \times 0 = 0$.

EXERCISES

1. Describe in words the following sets of numbers:
   (a) $N \cap Z$,   (b) $N \cap Z'$,
   (c) $Q' \cap R$,   (d) $Z' \cap Q$,
   (e) $N \cup (Q' \cap R)$.

2. Determine which of the following rational numbers are equal:
   (a) $\dfrac{111}{441}$ and $\dfrac{74}{294}$,   (b) $2.\overline{3}$ and $\dfrac{23}{10}$,
   (c) $.\overline{34}$ and $.3\overline{43}$,   (d) $.\overline{3}$ and $.\overline{33}$.

3. Express the following rational numbers in the form p/q, in which p and q have no common factors:
   (a) $2/3 + 3/7$,   (b) $1/2 - 1/6$,
   (c) $1.4 - 3/5$,   (d) $.8 + 1.5$,
   (e) $1.4\overline{14}$,   (f) $.5 - .\overline{005}$,
   (g) $.37 - .\overline{37}$,   (h) $1/4 + 1/5 + 1/6$,
   (i) $.\overline{11}/.\overline{22}$,   (j) $(.01 + .\overline{001})^2$,
   (k) $(3/4 - 7/8) \div 2/3$,   (l) $(2/3 - 1/2) \div 7/8$,
   (m) $2/3 \div (1/2 \div 7/8)$,   (n) $(2/3 - 1/2) - 3/7$,
   (o) $2/3 - (1/2 - 3/7)$,   (p) $(4/9(1/5 + 2/7)) \div 9/10$.

4. Using Theorem 3.1, show that

$$\sqrt{2} + 2/3 \in R \cap Q',$$

   i.e., that $\sqrt{2} + 2/3$ is irrational.

5. Show that $\sqrt{3}$ is an irrational number, i.e., $\sqrt{3} \in R \cap Q'$.

6. Express the following rational numbers as decimals, using the bar notation where necessary:
   (a) $2/3$,   (b) $.7/.8$,
   (c) $(1 - .05)/.01$,   (d) $.05/.11$,
   (e) $(1 + .05)^2$.

7. Write the appropriate inequalities, $a < b$ or $a > b$, that apply for the following pairs of numbers:
   (a) $a = 1.3, b = 1.\overline{3}$;   (b) $a = \frac{1}{2}, b = .55$;
   (c) $a = \sqrt{2}, b = 1.\overline{41}$;   (d) $a = \sqrt{2}, b = 1.4\overline{1}$;
   (e) $a = 6, b = x^2$, where $x = 1 + \sqrt{2}$;
   (f) $a = \sqrt{xy}, b = \frac{1}{2}(x + y)$, where $x = \sqrt{2}$ and $y = \sqrt{8}$;
   (g) $a = \sqrt{2} + \sqrt{3}, b = 3.1$;
   (h) $a = \sqrt{3} - \sqrt{2}, b = .\overline{3}$;
   (i) $a = 1/x, b = x$, where $0 < x < 1$;
   (j) $a = 1/x, b = x$, where $x > 1$.

1.4   Elementary Properties of Real Numbers

In the present section, we concentrate our attention on elementary properties of the various sets of numbers introduced in the preceding section. Our main goal here will be to increase the reader's manipulative skills in handling algebraic expressions. We will not place an excessive emphasis on the formal study of underlying axioms, although these certainly will be clearly set out.

Elementary algebra is concerned with the properties of numbers, operations on numbers, and relations between numbers. The numbers can come from any of the sets N, Z, Q, or R that were discussed in Section 1.3. The symbols that occur in elementary algebra are of two kinds: *constants* and *variables*. For example, "$+$," "-3," "$\div$" are typical constants. Variables are ordinarily indicated with letters (not necessarily English), e.g., "x," "y," "ω," "u," "t," "v," "a," "α," etc. A *formula* is an expression involving constants and variables which becomes a statement that is verifiably true or false upon replacement of the variables by numbers from some set. Thus

$$3x^2 + 5y - 17 = 28z + 5$$

is a formula. If x and y are replaced by 0 and z is replaced by $-11/14$, then the formula becomes a true statement. On the other hand, if x, y, and z are replaced by 0, the formula becomes the false statement

$$-17 = 5.$$

Many formulas in elementary algebra are preceded by *quantifiers*. A quantifier is a statement such as "for every" or "there exists." For example, the *commutative* law for addition in N can be stated:

For every x and y in N, $x + y = y + x$.

The existence of an *additive identity* in Z can be phrased: There exists an integer 0 such that

$$x + 0 = x$$

for any x in Z. A quantifier such as "for every $x \ldots$" is called *universal* whereas quantifiers of the type "there exists an element $\ldots$" are called *existential*. It is customary to omit quantifiers when the meaning is clear. Thus the formula

(1) $$3x + 5x + 7y + 2y = 8x + 9y$$

is true for all x and y and it is acceptable to write (1) without the usual prefatory "for every x and every y . . . ."

In an expression such as

$$3x + 8y - 10x^2z + 17,$$

each of the parts $3x$, $8y$, $10x^2z$, and 17 is called a *term*. In a product such as $10x^2z$, 10 is called the *coefficient* of x^2z. However, any one of the factors can also be referred to as a coefficient of the remaining factors. Hence in $10x^2z$, $10x^2$ is the coefficient of z, $10z$ is the coefficient of x^2, x^2 is the coefficient of $10z$, and z is the coefficient of $10x^2$.

An important part of elementary algebra is devoted to establishing certain universally true formulas. For example, it is clear that the following statement holds: For all numbers x and y,

$$(2) \qquad\qquad 3x(2x + 7y) = 6x^2 + 21xy.$$

It is important to gain experience in verifying such formulas as (2). The reasoning behind the manipulation of formulas is quite simple. We permit just those operations on formulas that will not alter their validity when the variables are replaced by any numbers for which the formulas have meaning.

We shall assume that the reader is familiar with the usual laws or axioms that govern any of the sets of numbers N, Z, Q, or R. However, we recapitulate them here for convenience of reference.

Axiom 1 (Associativity) For all numbers r, s, and t,

$$(3) \qquad\qquad r + (s + t) = (r + s) + t$$

and

$$(4) \qquad\qquad r(st) = (rs)t.$$

Axiom II (Commutativity) For all numbers r and s,

$$(5) \qquad\qquad r + s = s + r$$

and

$$(6) \qquad\qquad rs = sr.$$

Axiom III (Distributivity)   For all numbers r, s, and t,

(7) $$r(s + t) = rs + rt$$

and

(8) $$(s + t)r = sr + tr.$$

Of course, (8) follows from commutativity of multiplication and (7). For,

$$r(s + t) = (s + t)r$$

and

$$rs = sr, \qquad rt = tr,$$

and hence (7) becomes (8).

For example, we can simplify the expression

(9) $$2x + \big(3(5x + 3y) + 8y\big)$$

by repeatedly using the above axioms. For, by (7) and (4),

$$3(5x + 3y) = 3(5x) + 3(3y)$$
$$= 15x + 9y$$

and then by (3),

$$\big(3(5x + 3y) + 8y\big) = \big((15x + 9y) + 8y\big)$$
$$= 15x + 17y.$$

Hence, using (3) once more, (9) becomes

$$2x + (15x + 17y) = 17x + 17y.$$

In general, we will not burden ourselves with a quotation of the axioms in simplifying expressions.

If in a product of several numbers all the numbers are equal, we use a shortened notation. Thus for example, instead of $a \cdot a \cdot a \cdot a \cdot a$, we write a^5 where the 5 indicates that there are five factors equal to a. It is remarkable that this simple notational device provides an important method of turning multiplication problems into addition problems, as we shall show in the last section of Chapter 2 on logarithmic computations.

Definition 4.1 (Integer Exponents) If a is a real number and n a positive integer, then

$$a^n = \underbrace{a \cdot a \cdot \ldots \cdot a}_{n \text{ factors}}.$$

If $a \neq 0$, then it is convenient to define

(10) $$a^0 = 1,$$

and

(11) $$a^{-n} = \frac{1}{a^n}.$$

The number a^n is called the nth *power* of a, and the numbers a and n are called the *base* and the *exponent* of a^n, respectively.

Example 4.1

(a) $a^3 = a \cdot a \cdot a,$
(b) $4^0 = 1,$
(c) $2^{-3} = 1/2^3 = 1/(2 \cdot 2 \cdot 2) = 1/8,$
(d) $(-1)^{-2} = 1/(-1)(-1) = 1/1 = 1,$
(e) $2^{2^2} = 2^4 = 2 \cdot 2 \cdot 2 \cdot 2 = 16.$

The following laws of exponents follow almost immediately from the definition:

(12) $$a^n a^m = a^{n+m},$$
(13) $$(a^n)^m = a^{nm},$$
(14) $$(ab)^n = a^n b^n,$$
(15) $$a^n/a^m = a^{n-m}, \quad a \neq 0,$$
(16) $$(a/b)^n = a^n/b^n, \quad b \neq 0.$$

For example, if m and n are positive integers, then

$$a^n a^m = \overbrace{a \cdots a}^{n} \cdot \overbrace{a \cdots a}^{m}$$
$$= \underbrace{a \cdots a}_{n+m}$$
$$= a^{n+m}.$$

If m is a positive integer and n is a negative integer, say $n = -k$, then

$$a^n a^m = \overbrace{a \cdots a}^{m} \cdot \dfrac{1}{\underbrace{a \cdots a}_{k}}$$

$$= \begin{cases} \overbrace{a \cdots a}^{m-k} & \text{if } m > k, \\ 1 & \text{if } m = k, \\ \dfrac{1}{\underbrace{a \cdots a}_{k-m}} & \text{if } m < k \end{cases}$$

$$= \begin{cases} a^{m-k} & \text{if } m > k, \\ a^{m-k} & \text{if } m = k, \\ a^{m-k} & \text{if } m < k \end{cases}$$

$$= a^{m-k}$$
$$= a^{m-(-n)}$$
$$= a^{m+n}$$
$$= a^{n+m}.$$

In case m is negative and n is positive or both are negative, the formula (12) is similarly established. Formulas (13), (14), (15), (16) are done in much the same way and are left as exercises.

Example 4.2   (a) Simplify $(3x^2y^5)^2$.
   From (14) and (13) in succession we have

$$(3x^2y^5)^2 = 3^2(x^2)^2(y^5)^2$$
$$= 9x^{2\cdot 2}y^{5\cdot 2}$$
$$= 9x^4 y^{10}.$$

   (b)   Simplify $(-3)^{-3}$.
   From (11),

$$(-3)^{-3} = 1/(-3)^3$$
$$= 1/((-3)\cdot(-3)\cdot(-3))$$
$$= -1/27.$$

We can, of course, write (if we prefer)

$$-\dfrac{1}{27} = (-27)^{-1}.$$

Definition 4.2 (Rational Exponents) If a is a positive real number and q a nonzero integer, then $a^{1/q}$ is the unique positive number which satisfies

$$(a^{1/q})^q = a.$$

We sometimes write

$$a^{1/q} = \sqrt[q]{a}$$

and call this the *principal* qth *root* of a, or simply the qth *root* of a; $\sqrt[q]{a}$ is also called a *radical*. In case $q = 2$ it is customary to omit the 2 in $\sqrt[2]{a}$, i.e., $a^{1/2} = \sqrt{a}$. Let p/q be a rational number, p and q integers, $q \neq 0$, and let a be a positive real number. We define

(17) $$a^{p/q} = (a^{1/q})^p.$$

Clearly if r is a nonzero integer, then

$$a^{pr/qr} = a^{p/q}.$$

For,

$$((a^{1/q})^{1/r})^{qr} = [((a^{1/q})^{1/r})^r]^q$$
$$= ((a^{1/q})^1)^q$$
$$= a,$$

and therefore, by definition, $(a^{1/q})^{1/r} = a^{1/qr}$. Thus

$$a^{pr/qr} = (a^{1/qr})^{pr}$$
$$= [((a^{1/q})^{1/r})^r]^p$$
$$= (a^{1/q})^p$$
$$= a^{p/q}.$$

Our notation is therefore consistent with the properties of fractions.

Example 4.3 The following equalities follow immediately from the definition:

$$4^{1/2} = 2,$$
$$8^{-1/3} = (8^{1/3})^{-1}$$
$$= 2^{-1}$$
$$= \tfrac{1}{2},$$
$$8^{2/3} = (8^{1/3})^2$$
$$= 2^2$$
$$= 4,$$

$$(.008)^{-2/3} = ((.008)^{1/3})^{-2}$$
$$= (.2)^{-2}$$
$$= (.04)^{-1}$$
$$= 25.$$

Theorem 4.1   *Let a and b be positive real numbers. Then for any rational numbers m and n and any integers p and q, q > 0:*

(a) $(a^n)^{-1} = a^{-n}$;

(b) $(a^p)^{1/q} = a^{p/q}$;

(c) $a^{m+n} = a^m a^n$;

(d) $(a^m)^n = a^{mn}$;

(e) $(ab)^n = a^n b^n$.

Proof   (a)   Let $n = p/q$, where p and q are integers, $q > 0$. Then

$$(a^n)^{-1} = (a^{p/q})^{-1}$$
$$= ((a^{1/q})^p)^{-1}$$
$$= (a^{1/q})^{-p},$$

by (13). Thus by (17),

$$(a^n)^{-1} = (a^{1/q})^{-p}$$
$$= a^{-p/q}$$
$$= a^{-n}.$$

(b)   We show that $a^{p/q}$ is the real qth root of a^p. By (17) and (13) we have

$$(a^{p/q})^q = ((a^{1/q})^p)^q$$
$$= (a^{1/q})^{pq}$$
$$= a^{pq/q}$$
$$= a^p.$$

(c)   Let $m = r/s$ and $n = p/q$, where r, s, p, q are integers, $s > 0, q > 0$. Then, using formula (17) of Definition 4.2 and formula (12) for integer exponents, we have

$$a^{m+n} = a^{r/s+p/q}$$
$$= a^{(rq+sp)/sq}$$
$$= (a^{1/sq})^{rq+sp}$$
$$= (a^{1/sq})^{rq}(a^{1/sq})^{sp}$$
$$= a^{rq/sq}a^{sp/sq}$$
$$= a^{r/s}a^{p/q}$$
$$= a^m a^n.$$

(d) We saw that for any positive real number b and any nonzero integers s and q,

(18) $$(b^{1/s})^{1/q} = b^{1/sq}.$$

Now, using the notation of the proof of (c), we have by (17), part (b), and (18),

$$\begin{aligned}
(a^m)^n &= (a^{r/s})^{p/q} \\
&= ((a^{1/s})^r)^{p/q} \\
&= (((a^{1/s})^r)^p)^{1/q} \\
&= ((a^{1/s})^{rp})^{1/q} \\
&= (a^{rp/s})^{1/q} \\
&= ((a^{rp})^{1/s})^{1/q} \\
&= (a^{rp})^{1/sq} \\
&= a^{rp/sq} \\
&= a^{mn}.
\end{aligned}$$

(e) If $n = p/q$ we have, by (14),

$$\begin{aligned}
(ab)^n &= (ab)^{p/q} \\
&= ((ab)^p)^{1/q} \\
&= (a^p b^p)^{1/q}.
\end{aligned}$$

It remains to show that

(19) $$(a^p b^p)^{1/q} = a^{p/q} b^{p/q},$$

i.e., that $a^{p/q} b^{p/q}$ is the qth root of $a^p b^p$. Now, by (14) and part (d),

$$\begin{aligned}
(a^{p/q} b^{p/q})^q &= (a^{p/q})^q (b^{p/q})^q \\
&= a^{pq/q} b^{pq/q} \\
&= a^p b^p,
\end{aligned}$$

and (19) is proved. ∎

Example 4.4

(a) $(27)^{2/3} = ((27)^{1/3})^2 = 3^2 = 9$;

(b) $1000^{-1/3} = ((1000)^{1/3})^{-1} = 10^{-1} = 1/10$;

(c) $(256)^{-1/4} = ((256)^{1/4})^{-1} = 4^{-1} = 1/4$;

(d) $\left(\dfrac{100x^2 y^4}{49z^8}\right)^{1/2} = \dfrac{(100x^2 y^4)^{1/2}}{(49z^8)^{1/2}} = \dfrac{100^{1/2}(x^2)^{1/2}(y^4)^{1/2}}{(49)^{1/2}(z^8)^{1/2}} = \dfrac{10xy^2}{7z^4}$;

(e) $\dfrac{2x^5y^3}{x^3y^2} = 2x^{5-3}y^{3-2} = 2x^2y;$

(f) $\sqrt{\dfrac{121x^6}{9y^4}} = \left(\dfrac{11^2x^6}{3^2y^4}\right)^{1/2} = \dfrac{(11^2)^{1/2}(x^6)^{1/2}}{(3^2)^{1/2}(y^4)^{1/2}} = \dfrac{11x^3}{3y^2}.$

SYNOPSIS

In this section we defined the following items: formula, universal and existential quantifiers, term, and coefficient. The basic laws or axioms are listed in (3) through (8). In (12) through (16) and Theorem 4.1, the laws of exponents are stated.

QUIZ

Answer *true* or *false:*

1. $x^2 + y^2 = 0$ is a formula.
2. "For every x and y, $x^2 + y^2 = 0$" is a statement involving a universal quantifier.
3. "There exists a circle of radius 5" is a statement involving a universal quantifier.
4. The coefficient of $3x^2y^3$ in $75x^5y^5$ is $25x^3y^2$.
5. The terms in the expression $2x^2 + 15x + 7$ are $2x^2$, $15x$, and 7.
6. $8 + 9 = 9 + 8$ is an example of the associative law.
7. $(2^0)^0 = 0.$
8. $(2^0)^2 = 1.$
9. $(2^2)^{1/2}$ can be either 2 or -2.
10. $(-1)^{-(1^{-1})} = 1.$

EXERCISES

1. Remove parentheses and combine terms:
    (a) $(x^2 + 3xy + 5) - y(x + (x^2/y))$,
    (b) $-7(x - 2) + 7(2 - x)$,
    (c) $r(s - t) + t(r - s) + s(t - r)$,
    (d) $(x + y)(x - y) + 2xy$,
    (e) $-(x^2 + y^2)(x - y) + x^3 + (y^3 - 3xy^2)$,
    (f) $(x + y)(x - y)(x + y) - x(x + y)$,
    (g) $x^2 + x(y - x) - xy$,
    (h) $(3x^2 - 2y^2 - 2xy)(3x^2 + 2xy - 2y^2)$.
2. Express each of the following in the form $a^{p/q}$ in which p and q are integers having no common factors, i.e., the fraction p/q is in lowest terms.
    (a) $a^{2/3} \cdot a^{1/3}$,   (b) $a^5 \cdot a^{-3}$,   (c) $a^{-5} \cdot a^3$,

(d) $a^{1/2} \cdot a^{-2/3}$,

(e) $\dfrac{2^{1/2}}{2^{3/4}}$,

(f) $\dfrac{2^{3/4} \cdot 4^{3/2}}{2^{1/8}}$,

(g) $\dfrac{16^{1/4} \cdot 8^{2/3}}{4^{-3/2}}$,

(h) $(1/2^{-1})^{-1}$,

(i) $\dfrac{(-1)^{-1}}{3^{2/3}}$.

3. Simplify each of the following:

(a) $27^{-1/3}$,

(b) $\left(\dfrac{x^4}{125y^{12}}\right)^{4/3}$,

(c) $\left(\dfrac{x + x^{-1} - 2}{x}\right)^{1/2}$,

(d) $\dfrac{x^2 + 2xy + y^2}{(x + y)^3}$,

(e) $\sqrt[3]{a^2}\sqrt[4]{16a^3}$,

(f) $(\sqrt[3]{a^4})^{-4/3}$,

(g) $8\sqrt[3]{5}/2\sqrt[3]{25}$.

4. (a) Write $1/\sqrt{2} + 1/\sqrt{3}$ as a single fraction not involving fractional exponents in the denominator.

[*Hint:*

$$1/\sqrt{2} = \sqrt{2}/(\sqrt{2} \cdot \sqrt{2}) = \sqrt{2}/2.$$

Similarly, $1/\sqrt{3} = \sqrt{3}/3$. Then

$$1/\sqrt{2} + 1/\sqrt{3} = \sqrt{2}/2 + \sqrt{3}/3 = \dfrac{3\sqrt{2} + 2\sqrt{3}}{6}.]$$

Write each of the following as a single fraction not involving fractional exponents in the denominator:

(b) $\dfrac{3}{2^{1/2}} + \dfrac{5}{4^{3/2}}$,

(c) $1/2 + 1/\sqrt{2} + (1/\sqrt{2})^2$,

(d) $\dfrac{1}{3^{1/2}} + \dfrac{2}{4^{1/2}} + \dfrac{3}{5^{1/2}}$,

(e) $2/\sqrt{2} + 3/\sqrt{3} + 1/2 - 1/\sqrt{5}$.

5. (a) Express $3/(\sqrt{5} - \sqrt{2})$ as a fraction with no radicals in the denominator. This process is called *rationalizing the denominator*.

[*Hint:* Multiplying the indicated fraction by

$$(\sqrt{5} + \sqrt{2})/(\sqrt{5} + \sqrt{2}) = 1$$

does not, of course, alter its value. We then have

$$\dfrac{3}{\sqrt{5} - \sqrt{2}} = \dfrac{3(\sqrt{5} + \sqrt{2})}{(\sqrt{5} - \sqrt{2})(\sqrt{5} + \sqrt{2})}$$

$$= \dfrac{3(\sqrt{5} + \sqrt{2})}{5 - 2}$$

$$= \sqrt{5} + \sqrt{2}.]$$

Rationalize the denominator of each of the following expressions:

(b) $\dfrac{1}{\sqrt{x} - \sqrt{x+1}}$,

(c) $\dfrac{1}{\sqrt{x} + \sqrt{y}}$,

(d) $\dfrac{\sqrt{2} - 3}{2\sqrt{2} + \sqrt{3}}$,

(e) $\dfrac{1}{2\sqrt{3} + 1}$,

(f) $\dfrac{1}{\sqrt{x+1} - \sqrt{2x+5}}$.

2

Functions

2.1 Functions

Sets in general, and in particular sets of numbers, are the raw materials of mathematics. One of the most important concepts in mathematics, combining and comparing these entities, is that of a *function*. The word "function" is used by mathematicians in a somewhat different sense from its use in ordinary English ("social function," etc.). A function in mathematics means a special kind of relation between two sets. To define a function we need three things:

(i)   a nonempty set X, called the *domain* of the function;
(ii) a nonempty set Y, called the *range* of the function;
(iii) a rule f that prescribes for each element of X exactly one element of Y.

Thus a function consists of two sets and a certain type of relation between these sets. We denote the function thus defined by $f: X \to Y$. If the domain and the range of $f: X \to Y$ are clearly understood from the context, we write f instead of $f: X \to Y$ and speak of the function f. If x is an element of the domain X and the function f assigns to x the element y of the range Y, we say that y is the *value* of f at x, and we write $y = f(x)$; the symbol $f(x)$ is read "f of x." We also say that the function f *maps* x *into* y. Two functions $f: X \to Y$ and $g: X \to Y$ are said to be *equal* if $f(x) = g(x)$ for all x in X. Note that equal functions must have the same domain and the same range, and they must map each element of the domain into the same element of the range. If W is a subset of the domain X of f, then the set of all values of $f(x)$ for x in W is denoted by $f(W)$ and is called the *image of W under f*. In particular, $f(X)$ is the set of all values of f; $f(X)$ is called the *image set* of f.

37

Example 1.1   Let Z denote the set of integers, E the set of even integers, and let $f: Z \to E$ be defined by the formula

$$f(z) = 2z$$

for all z in Z. Then, for example, $f(0) = 0, f(1) = 2, f(2) = 4, f(-5) = -10$. It is easy to see that $f(Z) = E$. Also, the function f has the property that for any y in the range of f, there exists exactly one z in Z such that $y = f(z)$. In fact, $z = \frac{1}{2}y$.

Example 1.2   Let X be the set of all workers in the United States, and let N be the set of natural numbers. Define a function $f: X \to N$ as follows: For any $x \in X$, $f(x)$ is the Social Security number of x. Obviously f is a function, and moreover, if x_1 and x_2 are two different people, then $f(x_1) \neq f(x_2)$, i.e., different people have different Social Security numbers.

Example 1.3   Let X be the set of all rectangles in a plane and let R be the set of real numbers. Define $f: X \to R$ as the function that associates with each rectangle the number of square inches in the area of the rectangle. Note that in this example $f(X) \neq R$. Indeed, $f(X)$ is the set of positive real numbers, a subset of R. Also, for each positive number y in R, there is more than one rectangle in X whose area is y square inches, i.e., more than one element in the domain X at which the value of f is exactly y. It turns out that these two characteristics of a function are of great interest. It is often very important to know if a given function maps distinct elements of its domain into distinct elements of its range and if every element of the range is in the image of the domain.

Definition 1.1   (One–one; Onto)   If $f: X \to Y$ is a function such that

$$f(x_1) = f(x_2)$$

always implies $x_1 = x_2$, then f is said to be *one–one*, sometimes written 1–1. In other words, a function is one–one if it has different values at different elements of the domain.

If $f: X \to Y$ is a function such that $f(X) = Y$ (i.e., for each y in Y there exists an x in X, not necessarily unique, satisfying $y = f(x)$), then the function f is said to be *onto* Y. We also say, somewhat ungrammatically, that f is an *onto function*.

The function in Example 1.1 is both 1–1 and onto. In Example 1.2 the function is 1–1 but clearly not onto. The function in Example 1.3 is neither 1–1 nor onto.

We now give some additional examples of functions. It is clear that if the domain of a function contains a large number of elements (or indeed infinitely many elements), it may not be possible to explicitly give the value of the function at each element of the domain. In such cases the value may be prescribed by some method for computing it or by describing a distinguishing property of the value at a given element.

Example 1.4 (a) Let X be any nonempty set and let c be any fixed element of X. Define a function $f: X \rightarrow X$ by the formula:

$$f(x) = c$$

for all x in X. That is, f is the function mapping every element of the domain into the same element of the range. This function is called a *constant* function. Clearly, if X contains more than one element (which is the case most often), then f is neither 1–1 nor onto.

(b) Let $X = \{a, b, c, d\}$ and $Y = \{u, v, w\}$, where a, b, c, d, u, v, w are distinct elements. Now let $f: X \rightarrow Y$ be defined by: $f(a) = v, f(b) = w, f(c) = u, f(d) = v$. The function f is not 1–1, for $f(a) = f(d)$ while $a \neq d$. However, f is an onto function; for, $u = f(c), v = f(d), w = f(b)$, and therefore $f(X) = Y$.

(c) Let $X = \{a, b, c, d\}$ and let $g: X \rightarrow X$ be defined by: $g(a) = c, g(b) = a$, $g(c) = b, g(d) = d$. In this example the domain and the range are the same set. The function g is clearly 1–1 and onto X. In general, any 1–1 function that maps a finite set S onto itself is called a *permutation* of S.

(d) Let S be any nonempty set. Define a function $I_S: S \rightarrow S$ by $I_S(x) = x$ for each x in S; that is, the function I_S maps each element of S into itself. Obviously I_S is 1–1 and onto. It is called the *identity function* of S.

(e) Let R be the set of all real numbers and let $F: R \rightarrow R$ be defined by the formula:

$$F(x) = \frac{x^2 - 2}{x^2 + 3}$$

for all x in R. The formula describes how to compute the value of F at any number x: Square x (multiply x by x), subtract 2, and then divide the resulting number by the number obtained by squaring x and adding 3. Thus:

$$F(4) = \frac{4 \times 4 - 2}{4 \times 4 + 3} = \frac{14}{19},$$

$$F(1) = \frac{1 \times 1 - 2}{1 \times 1 + 3} = -\frac{1}{4},$$

$$F(0) = \frac{0 \times 0 - 2}{0 \times 0 + 3} = -\frac{2}{3},$$

$$F(-4) = \frac{(-4)(-4) - 2}{(-4)(-4) + 3} = \frac{14}{19}.$$

We observe that $F(4) = F(-4)$, and therefore F is not 1–1. We also note that $F(x)$ cannot be equal to 1 for any x. For, if such an x existed, then for this x we would have

$$\frac{x^2 - 2}{x^2 + 3} = 1$$

and therefore

$$x^2 - 2 = x^2 + 3$$

which is impossible. Hence F is not onto.

If $f: X \to Y$ is a 1–1 function onto Y, then the function f pairs off the elements of X with those of Y in such a way that to each element of X there corresponds exactly one element of Y and vice versa. (We describe this situation by saying that f establishes a *one–one correspondence* between X and Y.) Given any element y of Y there exists a unique element x_y in X such that

(1) $$f(x_y) = y.$$

The existence and uniqueness of x_y are guaranteed by the fact that f is onto and 1–1, respectively. We can thus define a function $g: Y \to X$ by

(2) $$g(y) = x_y$$

for all y in Y.

Definition 1.2 (Inverse Function) If $f: X \to Y$ is 1–1 onto Y, then the function $g: Y \to X$ uniquely defined by (1) and (2) is called the *inverse function* of f. The function g is denoted by f^{-1}.

Example 1.5 (a) Let $f: R \to R$ be defined by

(3) $$f(x) = 3x + 2.$$

Find the inverse function f^{-1}.

Clearly f is 1–1 onto R. We want to find a rule that will associate with each y a number x_y satisfying $y = f(x_y)$, i.e.,

(4)
$$y = 3x_y + 2.$$

We have $f^{-1}(y) = x_y$; we solve (4) for x_y to obtain

$$x_y = \tfrac{1}{3}y - \tfrac{2}{3},$$

i.e.,

(5)
$$f^{-1}(y) = \tfrac{1}{3}y - \tfrac{2}{3}.$$

Formula (5) defines the inverse function f^{-1}. In some cases it may be convenient to denote the general element of the domain of f^{-1} by x rather than y. We can rewrite (5) as

$$f^{-1}(x) = \tfrac{1}{3}x - \tfrac{2}{3}.$$

(b) Let σ be a permutation on the set $N_3 = \{1, 2, 3\}$ given by

$$\sigma(1) = 3, \qquad \sigma(2) = 1, \qquad \sigma(3) = 2.$$

By the definition of a permutation, the inverse σ^{-1} exists and from Definition 1.2 we immediately have

$$\sigma^{-1}(1) = 2, \qquad \sigma^{-1}(2) = 3, \qquad \sigma^{-1}(3) = 1.$$

SYNOPSIS

In this section we defined a function from X to Y, $f: X \rightarrow Y$. We introduced the concepts of a 1–1 function and an onto function. We gave examples of functions and defined a constant function, a permutation, the identity function, and the inverse of a 1–1 onto function.

QUIZ

Answer *true* or *false:*
1. If a function is 1–1, then it is onto.
2. If a function is onto, then it is 1–1.
3. If f is a 1–1 function from a finite set X to X, then f is onto X.
4. If X is a finite set and f is a function from X onto X, then f is 1–1.
5. If f is a function from S to itself and if f is not 1–1, then f cannot be onto S.
6. If $x_1 = x_2$ implies $f(x_1) = f(x_2)$, then f is a 1–1 function.
7. Let $N_2 = \{1, 2\}$. Then there are four distinct functions from N_2 to N_2.
8. There is only one permutation of the set $N_2 = \{1, 2\}$.

9. If P is the set of positive real numbers, and g is the function from P to itself defined by $g(x) = \dfrac{1}{x}$ for all x in P, then $g^{-1} = g$.

10. If $f: R \to R$ is the function given by the formula $f(x) = 3$ for all x in R, then f^{-1} is defined by $f^{-1}(x) = \frac{1}{3}$ for all x in R.

EXERCISES

1. Let $f: R \to R$ be defined by

$$f(x) = 3x^2 + 6x - 2$$

for all $x \in R$. Evaluate $f(-2), f(-1), f(0), f(1),$ and $f(2)$. Is f one–one?

2. Let $t: Z \to Z$, where Z is the set of integers, be defined by

$$t(z) = -z$$

for all $z \in Z$. Is t one–one? Is t onto Z?

3. Find the image $f(X)$ if X is the stated subset of R and $f: X \to R$ is defined by:
   (a) $f(x) = 4$ for all $x \in X$, $X = R$;
   (b) $f(x) = 2x - 5$ for all $x \in X$, $X = R$;
   (c) $f(x) = 1/x$ for all $x \in X$, $X = \{x \mid 1 \le x \le 2\}$;
   (d) $f(x) = x^2$ for all $x \in X$, $X = R$;
   (e) $f(x) = x^2 - 3$ for all $x \in X$, $X = R$;
   (f) $f(x) = \begin{cases} 2x, & \text{for all } x \le 2, \\ 6 - x, & \text{for all } x \ge 2, \end{cases} \quad X = R.$

4. Let $N_3 = \{1, 2, 3\}$.
   (a) Write out all constant functions from N_3 to N_3.
   (b) Write out all permutations of N_3.
   (c) Define a function from N_3 to N_3 that is neither a permutation nor a constant function.

5. Let $f: R \to R$ be a function from the set of real numbers to itself and let $f(x)$ denote, as usual, the value of f at x. Let $f(x)$ be given by the formula

$$f(x) = \frac{1}{1 + x}$$

for all x except $x = -1$, and let

$$f(-1) = 0.$$

Show that f is 1–1 and onto.

6. Let $f: R \rightarrow R$ be given by the following formulas. Find a formula for $f^{-1}(x)$ for each of these functions:

(a) $f(x) = x$;

(b) $f(x) = 3x$;

(c) $f(x) = 3x - 2$;

(d) $f(x) = 2 - x$;

(e) $f(x) = 3 - 2x$;

(f) $f(x) = x^3$;

(g) $f(x) = \begin{cases} x + 2, & \text{if } x \leq 1, \\ 2x + 1, & \text{if } x \geq 1; \end{cases}$

(h) $f(x) = \begin{cases} 1/x, & \text{if } x \neq 0, \\ 0, & \text{if } x = 0. \end{cases}$

[*Hint:* See Example 1.5.]

2.2 Graphs of Functions

A function $f: X \rightarrow Y$ is defined if each element x of the domain X is associated with a unique element $f(x)$ of the range Y. In other words, a function determines and is determined by a set of pairs $(x, f(x))$ such that each element x of the domain X occurs in exactly one such pair. We say that the set of pairs is associated with the function f, or simply that it coincides with f. For example, the function in Example 1.4(b) is associated with the set of pairs $\{(a, v), (b, w), (c, u), (d, v)\}$; the function in Example 1.5(a) is associated with the infinite set of pairs

$$\{(x, y) \mid y = 3x + 2, x \in R\}.$$

If f happens to be a function from R to R (or from a subset of R to a subset of R), then the associated set of pairs is a set of pairs of real numbers. In this case it is possible to obtain an important geometric representation of f.

We saw in Section 1.3 that a real number can be represented uniquely as a point on a line. Once the number line has been defined, it is possible to set coordinates in the plane in a familiar way. To each pair of real numbers (a, b), it is possible to assign a unique point in the plane by constructing a pair of mutually perpendicular lines called *coordinate axes* intersecting at the *origin* O and by assigning to (a, b) that point P whose projections on the horizontal and vertical axes are a and b, respectively (Figure 2.1).

Figure 2.1

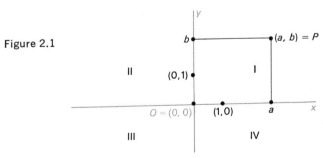

The coordinate axes divide the plane into four *quadrants*. We label these quadrants I, II, III, IV as shown in Figure 2.1. The four quadrants are characterized by the following inequalities:

$$
\begin{array}{ll}
\text{I:} & a \geq 0, b \geq 0, \\
\text{II:} & a \leq 0, b \geq 0, \\
\text{III:} & a \leq 0, b \leq 0, \\
\text{IV:} & a \geq 0, b \leq 0.
\end{array}
$$

(1)

Definition 2.1 (Graph) The *graph* of a function $f \colon X \to Y$, where X and Y are sets of real numbers, is the set of all points whose coordinates (x, y) satisfy

$$y = f(x).$$

Thus the graph of the function in Example 1.5(a) consists of the points on the straight line AB (Figure 2.2).

Figure 2.2

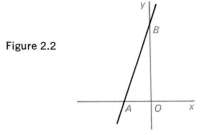

It can be shown that if $f(x)$ is given by a formula involving only addition, subtraction, multiplication, and division, then the graph of f is a smooth continuous curve, or else consists of several smooth continuous curves. The proof of this statement (and, indeed, the precise definition of "smooth continuous curve") is beyond the scope of this book.

Obviously it is quite impossible to accurately draw any but the simplest graphs. In order to sketch a graph, we plot a number of points of the graph, and, unless we have reason to suppose that the graph has discontinuities or "kinks," we connect these points by a continuous smooth curve. It is sometimes possible to use calculus to obtain further information about the shape of the graph, but, even with calculus, graph tracing is a rather approximate business. Although graphs play an important part in all experimental sciences, their use in mathematics is often overemphasized. It is true that we cannot prove theorems by sketching graphs. Graphs, however, are useful tools in studying functions. They enable us to visualize the behavior of functions, to suggest conjectures about functions, and to discard conjectures which are not likely to be true.

Example 2.1   (a)   Sketch the graph of the function $f: R \to R$ defined by

$$f(x) = 2 + 2x - x^2$$

for all x.

We compute:

$$
\begin{aligned}
f(-3) &= 2 + 2(-3) - (-3)^2 = -13,\\
f(-2) &= 2 + 2(-2) - (-2)^2 = -6,\\
f(-1) &= 2 + 2(-1) - (-1)^2 = -1,\\
f(0) &= 2 + 2 \cdot 0 - 0^2 = 2,\\
f(1) &= 2 + 2 \cdot 1 - 1^2 = 3,\\
f(2) &= 2 + 2 \cdot 2 - 2^2 = 2,\\
f(3) &= 2 + 2 \cdot 3 - 3^2 = -1,\\
f(4) &= 2 + 2 \cdot 4 - 4^2 = -6,\\
f(5) &= 2 + 2 \cdot 5 - 5^2 = -13.
\end{aligned}
$$

Observe that the function seems to grow nicely for values of x up to $x = 1$, and then to decrease continually. It may be worthwhile to examine the behavior of f for a few values of x near 1:

$$
\begin{aligned}
f(0.5) &= 2 + 2(0.5) - (0.5)^2 = 2.75,\\
f(1.5) &= 2 + 2(1.5) - (1.5)^2 = 2.75.
\end{aligned}
$$

We can also deduce, partly by computation, partly by intuition, that for numerically large negative values of x, $f(x)$ has numerically large negative values. Similarly, for large positive values of x, $f(x)$ has numerically large negative values. We shall not even attempt to define the meaning of the word "large," since the only purpose of the preceding comment is to obtain a rough idea of the shape of our graph beyond the computed values. We enter all the information in a table for easy reference.

x	$\to -\infty$	-3	-2	-1	0	0.5	1	1.5	2	3	4	5	$\to +\infty$
y	$\to -\infty$	-13	-6	-1	2	2.75	3	2.75	2	-1	-6	-13	$\to -\infty$

The symbols "$\to -\infty$" and "$\to +\infty$" are merely shorthand notations for "numerically large negative values" and "large positive values," respectively. We plot the points given in the table and connect them by a smooth curve (Figure 2.3). Using the graph, we may give an estimate of the value of the function at a point not included in the table, and, in particular, an estimate of the *zeros* of f, i.e., of the numbers x_i satisfying $f(x_i) = 0$. It is clear from the graph (Figure 2.3) that the

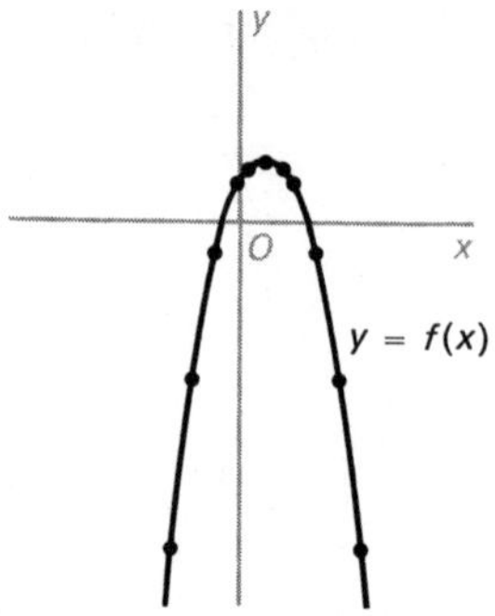

Figure 2.3

function f has two zeros: $x_1 = -.75$ and $x_2 = 2.75$, approximately. We can actually verify that

$$
\begin{aligned}
f(2.75) &= 2 + 2(2.75) - (-2.75)^2 \\
&= 2 + 5.50 - 7.56 \ldots \\
&= -.06 \ldots .
\end{aligned}
$$

Similarly, we can check that $f(-.75)$ is close to 0.

(b)   Let the function $g: R \to R$ be defined by

$$
g(x) = \tfrac{1}{3}(x^3 - 3x^2 - x + 3)
$$

for all x in R.   Sketch the graph of g.

We first compute the values of g at $-3, -2, -1, 0, 1, 2, 3, 4, 5$.

x	-3	-2	-1	0	1	2	3	4	5
y	-16	-5	0	1	0	-1	0	5	16

The values of g at numbers less than -3 or larger than 5 are numerically too large to be of use in tracing the graph.   Clearly $g(x) \to -\infty$ for $x \to -\infty$, and $g(x) \to \infty$ for $x \to \infty$.   It is impossible to sketch the graph of g from the above table, since these data do not contain enough information about the shape of the graph for x between -2 and 4.   We compute additional values

x	$-1\tfrac{1}{2}$	$-\tfrac{1}{2}$	$\tfrac{1}{2}$	$1\tfrac{1}{2}$	$2\tfrac{1}{2}$	$3\tfrac{1}{2}$
y	$-1\tfrac{7}{8}$	$\tfrac{7}{8}$	$\tfrac{5}{8}$	$-\tfrac{5}{8}$	$-\tfrac{7}{8}$	$1\tfrac{7}{8}$

and sketch the graph of g (Figure 2.4).

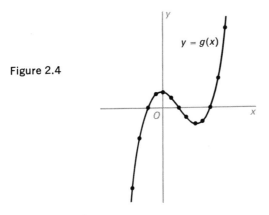

Figure 2.4

If a graph of a function is a straight line or if it consists of segments of straight lines, then it can be easily constructed, and the behavior of the function is apparent from the graph. We conclude this present section with a theorem which gives a necessary and sufficient condition for a graph of a function to be a straight line. The proof is based on elementary geometry; we assume only the properties of straight lines and similar triangles.

Theorem 2.1 *The graph of a function $f: R \to R$, from the set of real numbers to itself, is a straight line not parallel to the y-axis if and only if there exist fixed numbers m and b such that*

$$f(x) = mx + b \tag{2}$$

for all x in R.

Proof We show that any three points of the graph of f are collinear (i.e., lie on the same straight line) and the coordinates of any point on a straight line satisfy an equation of the form (2).

Let $x_1 < x_2 < x_3$, and let A, B, C be the points $(x_1, f(x_1))$, $(x_2, f(x_2))$, $(x_3, f(x_3))$, respectively, where

$$f(x_1) = mx_1 + b, \tag{3}$$

$$f(x_2) = mx_2 + b, \tag{4}$$

$$f(x_3) = mx_3 + b. \tag{5}$$

Draw AN parallel to the x-axis, BM and CN perpendicular to AN. We shall show that triangles AMB and ANC are similar, and it will follow that A, B, and C are collinear. Since the two triangles are right-angled, it suffices to prove that

$$\frac{MB}{AM} = \frac{NC}{AN}.$$

Figure 2.5

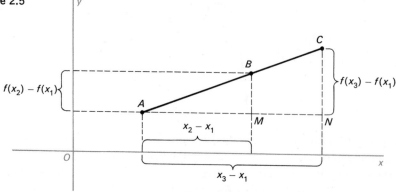

Now, subtracting (3) from (4) and from (5), we obtain

(6) $$f(x_2) - f(x_1) = m(x_2 - x_1),$$

(7) $$f(x_3) - f(x_1) = m(x_3 - x_1).$$

Hence, dividing both sides of (6) by $x_2 - x_1$ and both sides of (7) by $x_3 - x_1$, we have

$$\frac{f(x_2) - f(x_1)}{x_2 - x_1} = \frac{f(x_3) - f(x_1)}{x_3 - x_1} = m,$$

i.e.,

$$\frac{MB}{AM} = \frac{NC}{AN}.$$

Note that if $m = 0$, then $f(x_1) = f(x_2) = f(x_3) = b$, and A, B, C lie on the line $y = b$. Thus, in any case, the three points are collinear. Now let $G = (x', y')$ and $H = (x'', y'')$ be two fixed points on a straight line not parallel to the y-axis (i.e., $x' \neq x''$). We show that there exist numbers m and b such that for any point $P = (x, y)$ on the line GH we have $y = mx + b$. Draw GN parallel to the x-axis and MP and NH perpendicular to GN. Then, by elementary geometry, we have

$$\frac{MP}{GM} = \frac{NH}{GN},$$

Figure 2.6

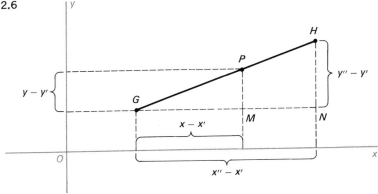

i.e.,

$$\frac{y - y'}{x - x'} = \frac{y'' - y'}{x'' - x'}.$$

Multiply both sides by $x - x'$:

$$y - y' = (x - x')\frac{y'' - y'}{x'' - x'},$$

and solve for y:

(8) $$y = \frac{y'' - y'}{x'' - x'}x + \left(y' - x'\frac{y'' - y'}{x'' - x'}\right).$$

Set

(9) $$m = \frac{y'' - y'}{x'' - x'},$$

(10) $$b = y' - x'\frac{y'' - y'}{x'' - x'}.$$

Then (8) becomes

(11) $$y = mx + b.$$

Now, m and b are fixed real numbers, independent of the coordinates x and y of P. It follows that any point (x, y) of the line through G and H satisfies equation (11). Thus all points on the line GH are on the graph of the function f defined by

$$f(x) = mx + b. \ \blacksquare$$

Example 2.2   Determine the function whose graph is the straight line through the points (1, 4) and (3, 2).

Here $(x', y') = (1, 4)$, $(x'', y'') = (3, 2)$ in (9) and (10) and we have

$$m = \frac{2 - 4}{3 - 1}$$
$$= -1$$

and

$$b = 4 - 1 \cdot \frac{2 - 4}{3 - 1}$$
$$= 5.$$

Therefore the required function is given by the formula

$$f(x) = -x + 5.$$

SYNOPSIS

Via coordinate geometry, we developed the representation of functions from reals to reals as graphs, that is, sets of points in a plane. In particular, the graph of a function $f: R \rightarrow R$ is a straight line not parallel to the y-axis if and only if

$$f(x) = mx + b$$

for some fixed numbers m and b and for all x in R.

QUIZ

Answer *true* or *false*:

1. If a, b, c are distinct real numbers, then the points (a, b), (b, a), (a, c) cannot belong to a graph of a function.
2. If a, b, c are distinct real numbers, then the points (a, b), (b, a), (c, a) cannot belong to a graph of a function.
3. Every straight line in a plane is a graph of some function.
4. If two functions $f: R \rightarrow R$ and $g: R \rightarrow R$ have the same graph, then they are equal.
5. The point $(-1, 5)$ lies in the third quadrant.
6. If $f: R \rightarrow R$ is defined by

$$f(x) = 3$$

for all x in R, then the graph of f is a straight line.

7. If $g: R \rightarrow R$ is defined by

$$g(x) = 3^2$$

for all x in R, then the graph of g is a straight line.
8. The graph of $f: R \rightarrow R$ must have a point on the y-axis.
9. The graph of $f: R \rightarrow R$ must have a point on the x-axis.
10. Let $f: R \rightarrow R$ and $g: R \rightarrow R$ be functions connected by the relation

$$f(x) = 3g(x) + 2$$

for all x in R. Then the graphs of f and g are both straight lines or neither of them is a straight line.

EXERCISES

1. Let $X = \{1, 2, 3, 4\}$, and let the permutation $\sigma: X \rightarrow X$ be defined by

$$\sigma(1) = 3, \qquad \sigma(2) = 4, \qquad \sigma(3) = 2, \qquad \sigma(4) = 1.$$

Plot the graph of σ.
2. Draw the graph of $f: R \rightarrow R$ defined by:
   (a) $f(x) = 4$;
   (b) $f(x) = -2$;
   (c) $f(x) = 0$;
   (d) $f(x) = x$;
   (e) $f(x) = -x$;
   (f) $f(x) = 3x - 2$;
   (g) $f(x) = 3x + 2$;
   (h) $f(x) = -3x + 2$;
   (i) $f(x) = 2x - 3$;
   (j) $f(x) = -2x + 3$;
   (k) $f(x) = \frac{1}{2}(x - 3)$;
   (l) $f(x) = \frac{2}{3}(\frac{1}{2} - x)$.
3. Draw the graph of $g: R \rightarrow R$ defined by

$$g(x) = \begin{cases} 2x + 1, & \text{if } x \le 2, \\ -x + 7, & \text{if } x > 2. \end{cases}$$

4. Draw the graph of $h: R \rightarrow R$ defined by

$$h(x) = \begin{cases} -x + 2, & \text{if } x < 3, \\ 2x - 4, & \text{if } x \ge 3. \end{cases}$$

5. Sketch the graph of $f: R \rightarrow R$, where f is defined by
   (a) $f(x) = \frac{1}{2}x^2$;
   (b) $f(x) = x^3 - x$;
   (c) $f(x) = \begin{cases} 1/x, & \text{if } x \ne 0, \\ 0, & \text{if } x = 0; \end{cases}$
   (d) $f(x) = \dfrac{4x}{x^2 + 1}$.

6. In each of the following, determine the function whose graph is the straight line through the two given points:

(a) (0, 0) and (1, 0),   (b) (1, 2) and (3, 2),
(c) (1, 2) and (2, 3),   (d) (−1, 2) and (2, 3),
(e) (1, −2) and (2, 3),   (f) (−1, −2) and (2, 3),
(g) (1, −2) and (2, −3),   (h) (1, −2) and (−2, −3),
(i) (7, 0) and (5, 1),   (j) (7, 1) and (5, 0),
(k) (−7, −1) and (−5, 0),   (l) (2/3, 1/3) and (1/3, 2/3),
(m) (2, 0) and (0, 5),   (n) (−3, 0) and (0, 1/2).

2.3 Lines and Distance

We saw in the preceding section that if $f: R \rightarrow R$ is a *linear* function, i.e., if

(1) $$f(x) = mx + b,$$

then its graph is a straight line l, not parallel to the y-axis.

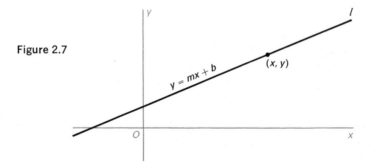

Figure 2.7

We say that $y = mx + b$ is the *equation of the line l*. In the event that a line is parallel to the y-axis, it consists of a set of points of the form (a, y), where a is some fixed number and y is arbitrary. We say $x = a$ is the equation of this line.

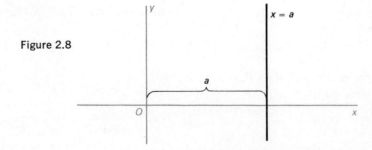

Figure 2.8

It is important in studying analytic geometry to have a formula that gives the distance between an arbitrary pair of points in the plane. In order to do this, we recall the famous theorem of Pythagoras from plane geometry:

Let $\triangle$ be a triangle with sides of length α, β, and γ. Then the equality

(2)
$$\gamma^2 = \alpha^2 + \beta^2$$

holds if and only if the sides whose lengths are α and β are perpendicular.

Figure 2.9

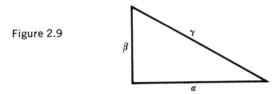

If equality holds in (2) then $\triangle$ is called a right-angled triangle or simply a *right triangle*. The Pythagorean theorem can be used to obtain a formula for the distance between any two points in the plane.

Figure 2.10

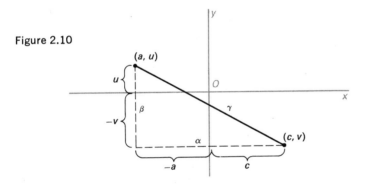

In Figure 2.10, the point (a, u) is in the second quadrant, and (c, v) is in the fourth quadrant. Hence, $a < 0$, $c > 0$, $u > 0$, $v < 0$ and thus $-a > 0$, and $-v > 0$. We have labeled the two segments that comprise the side of length α with their lengths, $-a$ and c; similarly we have labeled the two segments that comprise the side of length β with their lengths, u and $-v$. Thus we see from Figure 2.10 that

$$\beta = u + (-v)$$
$$= u - v,$$

and

$$\alpha = -a + c$$
$$= c - a.$$

The triangle in Figure 2.10 is a right triangle and thus

(3)
$$\gamma^2 = \alpha^2 + \beta^2$$
$$= (c - a)^2 + (u - v)^2$$
$$= (c - a)^2 + (v - u)^2.$$

Of course the points (a, u) and (c, v) can be situated differently than indicated in Figure 2.10. However, it is a routine matter to verify (see Exercise 14) that formula (3) is valid under any circumstances. Thus we have the following very important result.

Theorem 3.1 (Distance Formula) *Let (a, u) and (c, v) be any two points in the plane. Then the distance between (a, u) and (c, v), i.e., the length γ of the line segment joining the two points, is*

(4)
$$\gamma = \sqrt{(c - a)^2 + (v - u)^2}.$$

Example 3.1 (a) Find the distance between $(1, 1)$ and $(-3, 4)$.
Substituting directly into (4), we have

$$\gamma = \sqrt{(1 - (-3))^2 + (1 - 4)^2}$$
$$= \sqrt{4^2 + (-3)^2}$$
$$= \sqrt{16 + 9}$$
$$= \sqrt{25}$$
$$= 5.$$

(b) Find the point on the line $x + y + 1 = 0$ nearest to $(2, 3)$.
Let (c, v) be an arbitrary point on the line, i.e., $c + v + 1 = 0$. By (4), the square of the distance from $(2, 3)$ to (c, v) is

$$\gamma^2 = (c - 2)^2 + (v - 3)^2.$$

But $c = -v - 1$ and therefore

$$\gamma^2 = (-v - 1 - 2)^2 + (v - 3)^2$$
$$= v^2 + 6v + 9 + v^2 - 6v + 9$$
$$= 2v^2 + 18.$$

Since v^2 is always nonnegative, the smallest value that γ^2 can have is 18, and this value is assumed when $v = 0$. But $c + v + 1 = 0$, and thus if $v = 0$ then $c = -1$. It follows that $(-1, 0)$ is the closest point on the line to $(2, 3)$, and the distance between these two points is $\gamma = 3\sqrt{2}$.

(c) Find the point on the curve $x^2 + y^2 - 1 = 0$ closest to $(0, 3)$.

Let (c, v) be an arbitrary point on the curve. Then the square of the distance from $(0, 3)$ to (c, v) is given by (4) as

$$\gamma^2 = (c - 0)^2 + (v - 3)^2$$
$$= c^2 + v^2 - 6v + 9.$$

Now, (c, v) is a point on the curve $x^2 + y^2 - 1 = 0$ and hence $c^2 + v^2 = 1$. Thus

$$\gamma^2 = 1 - 6v + 9$$
$$= 10 - 6v.$$

Since $c^2 + v^2 = 1$, it follows that $v^2 = 1 - c^2 \leq 1$. Therefore $-1 \leq v \leq 1$, and hence, 6 is the largest value that $6v$ can assume when (c, v) is on the curve. Thus the smallest value that γ^2 can have is $10 - 6 = 4$, and this value is assumed when $v = 1$. But if $v = 1$, then $c^2 + v^2 = 1$ implies $c = 0$. Consequently, $(0, 1)$ is the closest point on the curve to $(0, 3)$, and the distance between these two points is $\gamma = 2$.

We shall apply Theorem 3.1 to determine conditions for two lines to be perpendicular. But before doing this, we must discuss some of the general properties of lines. Consider the equation of a line

(5)
$$y = mx + b.$$

If we set $x = 0$ in (5), we obtain b as the corresponding value of y. In other words, $(0, b)$ lies on the line. The number b is called the *y-intercept* of the line. Its numerical value is the distance from the origin to the point of intersection of the line with the y-axis. Similarly, if we set $y = 0$ in (5), we see that if $m \neq 0$ then $x = -b/m$. In other words, the point $(-b/m, 0)$ is on the line and the number $-b/m$ is called

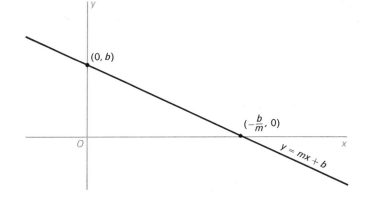

Figure 2.11

the *x-intercept*. Its numerical value is the distance from the origin to the point of intersection of the line with the *x*-axis. (Of course, if $m = 0$ then the equation (5) of the line becomes $y = b$. In other words, the line just consists of the totality of points (x, b), where x is arbitrary.) The general situation is described in Figure 2.11. The number m is called the *slope* of the line (5). We shall see immediately that the slope of a line can easily be computed once we know any two points on the line. In fact, we have the following result.

Theorem 3.2   *If (a, u) and (c, v) are any two points on the line $y = mx + b$, then the slope m is given by*

(6)
$$m = \frac{v - u}{c - a}.$$

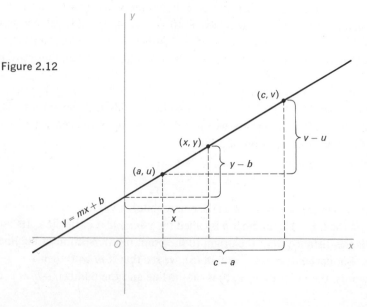

Figure 2.12

Proof   Since (a, u) and (c, v) are two different points on the line, it follows that $c \neq a$. (Why? See Exercise 15.) Then

$$v = mc + b,$$
$$u = ma + b,$$

and subtracting the second equation from the first we have

$$(v - u) = m(c - a).$$

Since $c - a \neq 0$, we can divide through the last equation to obtain formula (6). ∎

Example 3.2 (a) Let l be the line going through the points $(2, 3)$ and $(-1, -6)$. Find the slope of l. Find the equation of the line in the form (5).

We compute immediately from (6) that the slope of l is

$$m = \frac{3 - (-6)}{2 - (-1)}$$
$$= 9/3$$
$$= 3.$$

Thus the equation of the line is

$$y = 3x + b.$$

In order to determine b, we use the fact that the point $(2, 3)$ lies on the line. Thus

$$3 = 3 \cdot 2 + b,$$

and

$$b = -3.$$

Hence the equation of l is

$$y = 3x - 3.$$

(b) Find the equation of the line with slope 2 and x-intercept -3.

Since $m = 2$, the line has an equation of the form

$$y = 2x + b.$$

To say that the x-intercept is -3 means that $(-3, 0)$ is a point on the line. Thus $0 = 2(-3) + b$, and $b = 6$. Therefore the equation of the line is

$$y = 2x + 6.$$

(c) Find the equation of the line through $(2, -3)$ parallel to the line $8x + 4y = 9$.

Before we solve this problem, we observe that two distinct lines (not parallel to the y-axis) are parallel if and only if they have the same slope. For suppose that l_1 and l_2 have equations $y = m_1x + b_1$ and $y = m_2x + b_2$. The two lines will intersect if and only if the equations

(7)
$$-m_1x + y = b_1$$
$$-m_2x + y = b_2$$

have exactly one solution for x and y, i.e., if and only if $m_1 \neq m_2$. Hence the two lines are parallel if and only if $m_1 = m_2$ (and $b_1 \neq b_2$ if the lines are to be distinct).

In the present case the line whose equation is $8x + 4y = 9$ has slope -2, and thus the required line has an equation of the form $y = -2x + b$. The point $(2, -3)$ lies on the line and thus $-3 = -2 \cdot 2 + b$, which yields $b = 1$. Hence the equation of the required line is $y = -2x + 1$.

Theorem 3.3   *Let l_1 and l_2 be two lines, neither of which is parallel to the y-axis. If m_1 and m_2 are the slopes of l_1 and l_2, respectively, then l_1 and l_2 are perpendicular if and only if*

(8)
$$m_1 m_2 = -1.$$

Proof   First observe that l_1 and l_2 must intersect if condition (8) holds. Let them intersect at some point (a, u). The general situation is depicted in Figure 2.13.

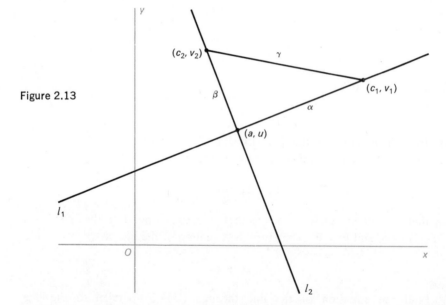

Figure 2.13

Let (c_1, v_1) be a point on l_1 and (c_2, v_2) a point on l_2. We compute the lengths α, β, and γ in Figure 2.13 by using the distance formula in Theorem 3.1:

(9)
$$\alpha^2 = (c_1 - a)^2 + (v_1 - u)^2,$$
$$\beta^2 = (c_2 - a)^2 + (v_2 - u)^2,$$
$$\gamma^2 = (c_2 - c_1)^2 + (v_2 - v_1)^2.$$

The lines l_1 and l_2 in Figure 2.13 are perpendicular if and only if the sides of lengths α and β are perpendicular. As we know from the Pythagorean theorem, a necessary and sufficient condition for this to happen is that

(10) $$\alpha^2 + \beta^2 = \gamma^2.$$

If we substitute the values (in (9)) of α^2, β^2, and γ^2, into the equality (10) and expand and cancel like terms, we obtain

(11) $$ac_1 - a^2 + uv_1 - u^2 + ac_2 + uv_2 = c_1c_2 + v_1v_2.$$

Rearranging and factoring the terms in (11), we obtain

(12) $$(a - c_2)(c_1 - a) = (v_2 - u)(v_1 - u).$$

(The student may also verify (12) by multiplying out both sides to obtain (11).) Now, $c_1 \neq a$ and $c_2 \neq a$, because we are assuming that neither l_1 nor l_2 is parallel to the y-axis. (See Figure 2.13.) Thus we can divide both sides of (12) by the negative of the left side of (12) to obtain

$$\frac{v_1 - u}{c_1 - a} \cdot \frac{v_2 - u}{c_2 - a} = -1.$$

But according to Theorem 3.2,

$$m_1 = \frac{v_1 - u}{c_1 - a}$$

and

$$m_2 = \frac{v_2 - u}{c_2 - a},$$

and thus (12) is precisely the statement (8). In other words, l_1 and l_2 are perpendicular if and only if the product of their slopes is -1. ∎

Example 3.3 (a) Find the equation of the line l_1 through the point $(0, 2)$ and perpendicular to the line l_2 whose equation is $2x + 3y = 5$.
    First rewrite the equation for the line l_2 in the form $y = (-2/3)x + 5/3$. Thus the slope m_2 of the line l_2 is

$$m_2 = -2/3.$$

For l_1 to be perpendicular to l_2, its slope m_1 must satisfy $m_1m_2 = -1$, i.e., $m_1(-2/3) = -1$ or $m_1 = 3/2$. Thus the equation of the line l_1 is of the form $y = (3/2)x + b$. Now, $(0, 2)$ is a point on l_1 and hence $2 = (3/2) \cdot 0 + b = b$. Thus l_1 has the equation $y = (3/2)x + 2$.

(b)   Show that if (a, u) and (c, v) are any two points, then the point

$$\left(\frac{a + c}{2}, \frac{u + v}{2}\right)$$

is the midpoint of the line segment joining the two points.

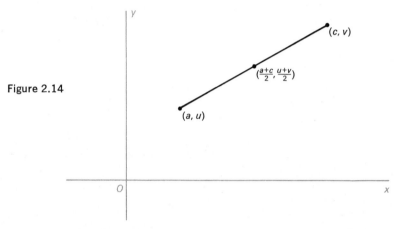

Figure 2.14

If $a = c$, then $\frac{1}{2}(a + c) = a$, and it is clear that the point $(a, \frac{1}{2}(u + v))$ is the midpoint of the vertical line segment joining (a, u) and (a, v). So assume that $a \neq c$. The equation of the line joining (a, u) and (c, v) is

(13) $$y - u = \frac{v - u}{c - a}(x - a).$$

(The student will verify this by putting the coordinates of (a, u) and (c, v) in (13) and seeing that the equation is satisfied.) Now, if we set $x = \frac{1}{2}(a + c)$ and $y = \frac{1}{2}(u + v)$ in (13), we obtain $\frac{1}{2}(v - u)$ on the left, and

$$\frac{v - u}{c - a} \cdot \frac{c - a}{2} = \frac{v - u}{2}$$

on the right.   Thus $\big((a + c)/2, (u + v)/2\big)$ is a point on the line joining (a, u) and (c, v).   The student will verify, using the distance formula, that $\big((a + c)/2, (u + v)/2\big)$ is in fact the same distance from (a, u) and (c, v).   (See Exercise 17.)

(c)   Find the equation of the line which is the perpendicular bisector of the line segment joining $(-1, 3)$ to $(5, 7)$.

According to the preceding example, the midpoint of the line segment is $\left(\dfrac{-1 + 5}{2}, \dfrac{3 + 7}{2}\right) = (2, 5)$.   The slope of the line joining $(-1, 3)$ and $(5, 7)$ is

$\left(\dfrac{7 - 3}{5 - (-1)}\right) = \dfrac{2}{3}$. Thus the slope of the perpendicular bisector is $-3/2$, so its equation is of the form $y = -(3/2)x + b$. But the perpendicular bisector must pass through $(2, 5)$ and hence $5 = -(3/2) \cdot 2 + b$, i.e., $b = 8$. Thus the required line has the equation $y = -(3/2)x + 8$.

SYNOPSIS

In Theorem 3.1 we derived the important formula (4) for the distance between two points in a plane. The notion of the slope of a line was introduced, and in Theorem 3.3 we obtained the criterion (8) for the perpendicularity of two lines in terms of their slopes. We then showed how to obtain the equations of lines satisfying various geometric conditions.

QUIZ

Answer *true* or *false:*
1. The equation of the x-axis is $y = 0$.
2. The distance between $(1, 1)$ and $(2, 2)$ is 1.
3. The point $(2, 3)$ bisects the segment joining $(5, 4)$ and $(-1, 5)$.
4. The slope of the line $8x + 5y - 3 = 0$ is -8.
5. Two lines with the same slope are parallel.
6. Two parallel lines have the same slope.
7. The lines $y = x$ and $y = -x$ are perpendicular.
8. The x-intercept of the line $y = x$ is 0.
9. The y-intercept of the line $y = -x$ is -1.
10. The point $(1, 1)$ is equally distant from $(0, 0)$ and $(2, 0)$.

EXERCISES

1. Find the distance between each of the following pairs of points:
    (a) $(1, 1)$, $(2, 2)$;
    (b) $(-1, 0)$, $(3, -4)$;
    (c) $(-3, 0)$, $(-1, -1)$;
    (d) $(2, 2)$, $(2, 2)$;
    (e) $(0, 0)$, $(0, -1)$;
    (f) $(4, 3)$, $(3, 4)$;
    (g) $(-1, -2)$, $(3, -4)$;
    (h) $(-2, 0)$, $(-3, 1)$;
    (i) $(1, 2)$, $(-2, 1)$;
    (j) $(0, 4)$, $(4, 0)$.
2. Find the coordinates of the midpoint of the line segment joining each of the following pairs of points. Graph the line segment and its midpoint.
    (a) $(1, 1)$, $(2, 3)$;
    (b) $(1, 1)$, $(3, -4)$;
    (c) $(2, 3)$, $(-1, -5)$;
    (d) $(3, 1)$, $(-1, 4)$;
    (e) $(-3, -2)$, $(-2, 5)$;
    (f) $(-3, 4)$, $(-2, -5)$.

3. For each of the following lines, find the x-intercept, the y-intercept, and the slope:
   (a) $y = 3x + 5$;
   (b) $-x - y = 1$;
   (c) $x/2 + y/3 = 1$;
   (d) $2x + 7y = 6$;
   (e) $-3y = 4x + 5$;
   (f) $x = y + 1$;
   (g) $5x + 9y = 11$;
   (h) $x + y = 0$;
   (i) $x - y = 0$;
   (j) $2x = y$;
   (k) $x = 2y$;
   (l) $3x + 6y = 9$.

4. Find the equation of the line which goes through each of the following pairs of points:
   (a) $(0, 0), (0, 1)$;
   (b) $(0, 0), (1, 0)$;
   (c) $(0, 0), (-1, 0)$;
   (d) $(2, 1), (1, 2)$;
   (e) $(3, 5), (-1, -1)$;
   (f) $(2, 2), (-2, -2)$;
   (g) $(3, 5), (-5, 0)$;
   (h) $(2, 1), (4, 6)$.

5. Find the equation of the line perpendicular to each of the following lines and going through the origin $(0, 0)$:
   (a) $y = x$;
   (b) $y = -x$;
   (c) $2x + 3y = 5$;
   (d) $2x + 3y = -5$;
   (e) $x + y = 1$;
   (f) $x - y = 1$;
   (g) $3x + 2y = 6$.

6. The equation $\alpha x + \beta y = d$ is said to be in *normal form* if $\alpha^2 + \beta^2 = 1$ and $d \geq 0$. Write each of the following equations in normal form:
   (a) $y - x = 1$;
   (b) $x - y = 1$;
   (c) $3x + 7y = 8$;
   (d) $x + y = 2$;
   (e) $x - y = -1$;
   (f) $2x - y = 3$;
   (g) $-x = y + 4$;
   (h) $y + 3x = 8x + 2$;
   (i) $2y = 7x - 3y$.

7. Show that if the equation of the line l is in normal form, $\alpha x + \beta y = d$, and $\beta \neq 0$, then any line l_1 perpendicular to l has an equation of the form $\beta x - \alpha y = c$. Show that if l_1 goes through the origin, then $c = 0$.

8. Using Exercise 7, find the equation of the line perpendicular to each of the following lines and going through the origin:
   (a) $x + y = 1$;
   (b) $x + 2y = 0$;
   (c) $3x - 4y = -1$;
   (d) $2x - 3y = 5$;
   (e) $x + y = -3x + 1$;
   (f) $2x + 2y = 8$;
   (g) $-y = 3x + 1$;
   (h) $2y - x = 0$;
   (i) $x - 2y = 0$;
   (j) $3y + 4x = -2$.

9. Find the equation of the line perpendicular to each of the following lines and going through the indicated points:
   (a) $2x + 3y = 5, (0, 0)$;
   (b) $-x - y = 1, (1, 2)$;
   (c) $x = 3, (1, 1)$;
   (d) $y = 1, (0, 0)$;
   (e) $x + y = 5, (-1, -2)$;
   (f) $2x + 3y = -1, (3, 2)$.

10. Let $\alpha x + \beta y = d$ be the equation of a line l in normal form (see Exercise 6). Assume that l is not parallel to either axis, and let (x_0, y_0) be an arbitrary point. Show that the distance from (x_0, y_0) to the line l is

$$\pm(\alpha x_0 + \beta y_0 - d).$$

The sign is so chosen that the value is nonnegative, i.e., it is a distance.
[*Hint:* See Figure 2.15. From Exercise 7, any line l_1 perpendicular to l has an equation of the form $\beta x - \alpha y = c$. If l_1 goes through the origin, then $c = 0$. The point of intersection of l and l_1 is $(\alpha d, \beta d)$, which is a distance d from the origin. In other words, d is the distance from l to the origin. The

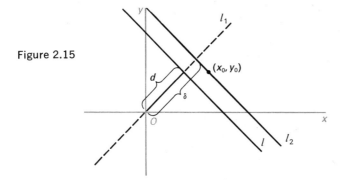

Figure 2.15

line l_2 through (x_0, y_0) and parallel to l has the equation $\alpha x + \beta y = \delta$, where δ is the distance from l_2 to the origin. The required distance is just the distance between the parallel lines l and l_2, i.e., it is

$$\delta - d = \pm(\alpha x_0 + \beta y_0 - d).]$$

11. In each of the following find the distance between the indicated points and lines. Remember to put the line in normal form before applying the formula in the preceding exercise.
    (a) $x + 2y = 1$, $(1, -1)$;   (b) $2x + 3y = 5$, $(0, 0)$;
    (c) $2y - x = 1$, $(4, -3)$;   (d) $x + 2y = 3y - 4x + 7$, $(2, 2)$;
    (e) $-x + 3y = -2$, $(-1, -3)$;   (f) $2x + y = -3$, $(-1, 4)$;
    (g) $x - 3y = 2$, $(2, -1)$;   (h) $2x + 3y = -x$, $(-1, -2)$;
    (i) $x + 5y = -9$, $(0, 5)$;   (j) $x/2 + y/3 = 1$, $(2, 7)$.
12. A *median* of a triangle is the line segment joining a vertex to the midpoint of the side opposite the vertex. Find the lengths of each of the three medians for the following triangles whose vertices are:
    (a) $(0, 0)$, $(0, 1)$, $(2, 2)$;   (b) $(3, 5)$, $(-1, -2)$, $(1, 2)$;
    (c) $(2, 3)$, $(-1, 0)$, $(1, 1)$;   (d) $(-1, -1)$, $(1, -2)$, $(0, 4)$;
    (e) $(0, 0)$, $(5, 3)$, $(3, 5)$.

13. For each of the triangles in the preceding exercise, find the equations of the lines containing each of the three medians. Show that these three lines meet in a single point.
    [*Hint:* Find the point of intersection of two of the medians and show that it lies on the third median.]

14. Show that if the points (a, u) and (c, v) are in the first and third quadrants, respectively, then the distance between them is given by formula (4). Draw the appropriate diagram analogous to Figure 2.10.

15. Suppose that (a, u) and (c, v) are two points on the line $y = mx + b$. Show that $c \neq a$.

16. Show that two perpendicular lines, neither of which is parallel to the y-axis, cannot have equal slopes.

17. Show that $(\frac{1}{2}(a + c), \frac{1}{2}(u + v))$ is equally distant from (a, u) and (c, v) by using the distance formula.

18. Find the equation of the line l satisfying the following geometric conditions:
    (a) parallel to the y-axis and with x-intercept 4;
    (b) with x-intercept 10 and parallel to the line $y - x = 0$;
    (c) equally distant from $(0, 1)$ and $(1, 0)$;
    (d) going through the point $(8, 7)$ and with slope $-2/3$;
    (e) going through the points $(-1, -1)$ and $(2, 4)$;
    (f) with slope -5 and y-intercept -10;
    (g) the perpendicular bisector of the line segment joining $(-3, 5)$ and $(2, -4)$.

19. Show that the quadrilateral (i.e., four-sided figure) with vertices $(-5, 7)$, $(-4, -6)$, $(8, 8)$, $(9, -5)$ is a square.

2.4 The Logarithmic Function

In Section 1.4 we extended the definition of powers of positive real numbers to powers with rational exponents. The next step is to define powers of positive real numbers with real exponents. Unfortunately, our intuitive approach to the set of real numbers (see the note immediately following Definition 3.1 in Chapter 1) does not allow us to give a rigorous definition of powers with irrational exponents nor to give a rigorous proof of an extension of Theorem 4.1 of Chapter 1 to powers of positive real numbers with real exponents x and y. Nevertheless, we shall attempt to explain the meaning of a^y if y is irrational.

We shall illustrate for $y = \sqrt{2}$ and $a = 3$. We proved in Chapter 1 (see Theorem 3.1) that $\sqrt{2}$ is irrational:

$$\sqrt{2} = 1.414213\ldots .$$

It is clear that if $y > z$ then $3^y > 3^z$. Thus we have an increasing sequence of powers whose rational exponents are all less than $\sqrt{2}$,

$$3^1 < 3^{1.4} < 3^{1.41} < 3^{1.414} < 3^{1.4142} < 3^{1.41421} < 3^{1.414213} < \cdots,$$

and a decreasing sequence of powers whose rational exponents are all greater than $\sqrt{2}$,

$$3^2 > 3^{1.5} > 3^{1.42} > 3^{1.415} > 3^{1.4143} > 3^{1.41422} > 3^{1.414214} > \cdots.$$

The terms in the second sequence are all greater than the terms in the first sequence, and the difference between the corresponding terms in the two sequences becomes progressively smaller. It can be shown that the difference between corresponding powers of 3 becomes as small as we wish, provided we take sufficiently close approximations of $\sqrt{2}$ for the exponents. There exists, therefore, a real number which is greater than any power in the first sequence and less than any power in the second sequence. It can also be shown that this number is uniquely determined. We naturally define it as $3^{\sqrt{2}}$. The real number $3^{\sqrt{2}}$ is therefore defined by a "nested" set of intervals

$$3^1 < 3^{\sqrt{2}} < 3^2,$$
$$3^{1.4} < 3^{\sqrt{2}} < 3^{1.5},$$
$$3^{1.41} < 3^{\sqrt{2}} < 3^{1.42},$$
$$3^{1.414} < 3^{\sqrt{2}} < 3^{1.415},$$

$$\cdot \quad \cdot \quad \cdot \quad \cdot \quad \cdot \quad \cdot \quad \cdot \quad \cdot$$

in which each interval is smaller than and is contained in the preceding one. By computing $\sqrt{2}$ to a sufficient number of decimal places, we can approximate $3^{\sqrt{2}}$ as closely as we wish by means of powers of 3 with rational exponents and the methods of this chapter.

In general, positive real numbers with irrational exponents can be defined in a similar way. It can also be shown that the analogue of Theorem 4.1 of Chapter 1 holds for all real exponents. Although we indicated how to define powers with irrational exponents in the preceding remarks, we have not shown how to compute $3^{\sqrt{2}}$ nor, in fact, how to compute a reasonable approximation of $3^{\sqrt{2}}$, e.g., $3^{1.41}$. A remarkable and dramatic consequence of the apparently abstract theory above is a simple and practical numerical method which allows us, *inter alia*, to obtain good approximations of real numbers such as $3^{\sqrt{2}}$.

Definition 4.1   (Exponential Function)   Let a be a positive real number, $a \neq 1$, and let R and P denote the set of real numbers and the set of positive real numbers, respectively. The function $\exp_a: R \to P$, defined by the formula

$$\exp_a(x) = a^x$$

for all x in R, is called the *exponential function to the base a.*

It follows immediately from the definition and from properties of powers that

$$\exp_a (x + y) = \exp_a (x) \cdot \exp_a (y),$$
$$\exp_a (x - y) = \exp_a (x)/\exp_a (y),$$
$$\left(\exp_a (x)\right)^y = \exp_a (xy),$$
$$\exp_a (1) = a,$$
$$\exp_a (0) = 1,$$

for all real numbers x and y. The preceding formulas are easily verified, e.g., $(\exp_a (x))^y = (a^x)^y = a^{xy}$ (see Theorem 4.1, Chapter 1).

The graph of $\exp_2$ is shown in Figure 2.16.

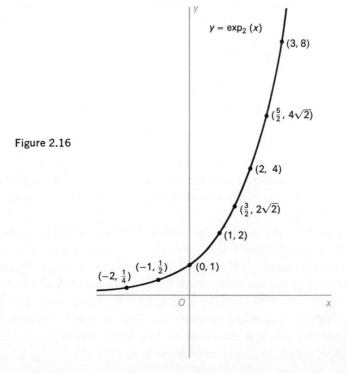

Figure 2.16

From the figure it appears that the function $\exp_2$ is 1–1 onto P. It can be shown* that the exponential function is always 1–1 onto, and thus it possesses an inverse.

*See, e.g., Marvin Marcus and Henryk Minc, *College Algebra*, Houghton Mifflin Co. (1970), pp. 225–228.

Definition 4.2 (Logarithmic Function) Let a be a positive real number $a \neq 1$. The inverse of the function $\exp_a : R \rightarrow P$ is called the *logarithmic function to the base a* and is denoted by

$$\log_a : P \rightarrow R.$$

If $x \in P$, then the value $\log_a (x)$ is called the *logarithm of x to the base a*. For simplicity we usually omit the parentheses and write $\log_a x$.

Since the logarithmic function to the base a is the inverse of the exponential function to the same base, any real numbers x and y that satisfy

$$\exp_a (y) = x$$

must also satisfy

$$\log_a x = y,$$

and vice versa. In other words, every positive real number x satisfies

$$\exp_a (\log_a x) = x,$$

and every real number y satisfies

$$\log_a \exp_a (y) = y.$$

Using the notation for powers, the two identities become

(1) $$a^{\log_a x} = x$$

for all x in P, and

(2) $$\log_a a^y = y$$

for all y in R. The identity (1) expresses the fact that the logarithm of x to the base a is the real number y such that $a^y = x$.

The graph of $\log_2$ together with that of $\exp_2$ is shown in Figure 2.17 on p. 68. The relation between the two graphs is clearly seen: If (α, β) is a point on one of the graphs, then (β, α) is a point on the other. We also note from the appearance of the two graphs that both functions are increasing for all numbers in their domains, i.e., that if x_1, x_2 are real numbers and $x_1 < x_2$ then

$$\exp_2 (x_1) < \exp_2 (x_2),$$

and if $0 < x_1 < x_2$ then

$$\log_2 x_1 < \log_2 x_2.$$

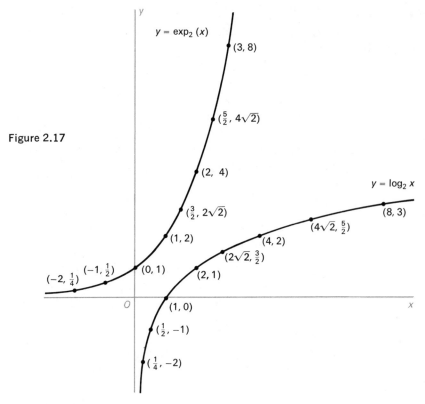

Figure 2.17

In fact it can be shown* that both $\exp_a$ and $\log_a$ are increasing functions for any $a > 1$.

Example 4.1 (a) Find the values of (*i*) $\log_{10} 100$, (*ii*) $\log_{10} 1$, (*iii*) $\log_{10} 0.01$. This is an easy and direct application of (2):

(*i*)
$$\log_{10} 100 = \log_{10} 10^2$$
$$= 2,$$

(*ii*)
$$\log_{10} 1 = \log_{10} 10^0$$
$$= 0,$$

(*iii*)
$$\log_{10} 0.01 = \log_{10} 10^{-2}$$
$$= -2.$$

(b) Solve for x: (*i*) $\log_2 x = 2$, (*ii*) $\log_2 x = 0$, (*iii*) $\log_2 x = -2$.

*See, e.g., Marcus and Minc, *College Algebra*, Houghton Mifflin Co., p. 223 and p. 235.

Using the definition of the function $\log_2$ (or using formula (1) with $a = 2$), we have

(i) $\qquad\qquad\qquad\qquad\qquad x = 2^2 = 4,$

(ii) $\qquad\qquad\qquad\qquad\qquad x = 2^0 = 1,$

(iii) $\qquad\qquad\qquad\qquad\qquad x = 2^{-2} = \frac{1}{4}.$

(c) Solve the following equations for a: (i) $\log_a 81 = 4$, (ii) $\log_a 81 = -4$. (i) Apply identity (1) with $x = 81$ and $\log_a x = 4$:

$$a^4 = 81$$

and hence

$$a = 3.$$

(ii) Here $x = 81$ and $\log_a x = -4$. Thus

$$a^{-4} = 81,$$
$$a^{-1} = 3,$$
$$a = \tfrac{1}{3}.$$

The following properties of the logarithmic function follow directly from the properties of the exponential function or from the corresponding properties of powers.

Theorem 4.1 *Let x, y, a, b, be positive real numbers, and assume that neither a nor b is equal to 1. Then*

(3) $\qquad\qquad\qquad \log_a (xy) = \log_a x + \log_a y,$

(4) $\qquad\qquad\qquad \log_a \dfrac{x}{y} = \log_a x - \log_a y,$

(5) $\qquad\qquad\qquad \log_a x^r = r \log_a x,$ *for any real number r,*

(6) $\qquad\qquad\qquad \log_a a = 1,$

(7) $\qquad\qquad\qquad \log_a 1 = 0,$

(8) $\qquad\qquad\qquad \log_b x = \dfrac{\log_a x}{\log_a b}.$

Proof   To prove (3) we use (1) and the properties of powers:

$$
\begin{aligned}
a^{\log_a(xy)} &= xy \\
&= a^{\log_a x} a^{\log_a y} \\
&= a^{\log_a x + \log_a y}.
\end{aligned}
$$

Since an exponential function is 1–1, the expressions in the exponents must be equal.   Similarly we have

$$
\begin{aligned}
a^{\log_a(x/y)} &= x/y \\
&= a^{\log_a x} / a^{\log_a y} \\
&= a^{\log_a x} a^{-\log_a y} \\
&= a^{\log_a x - \log_a y},
\end{aligned}
$$

which implies (4).

Next we compute

$$
\begin{aligned}
a^{\log_a x^r} &= x^r \\
&= (a^{\log_a x})^r \\
&= a^{r \log_a x},
\end{aligned}
$$

and (5) follows.

Identities (6) and (7) follow directly from the definition of the logarithmic function.   Lastly we prove (8).   We have

$$
x = b^{\log_b x} \tag{9}
$$

and

$$
x = a^{\log_a x}. \tag{10}
$$

Now from (1)

$$
b = a^{\log_a b},
$$

and therefore (9) yields

$$
\begin{aligned}
x &= (a^{\log_a b})^{\log_b x} \\
&= a^{(\log_a b)(\log_b x)}.
\end{aligned} \tag{11}
$$

Therefore from (10) and (11),

$$
\log_a x = (\log_a b)(\log_b x),
$$

which is equivalent to (8), since $b \neq 1$, and therefore $\log_a b \neq 0$.   ∎

Example 4.2 (a) Show that

(12) $$\log_b a = 1/\log_a b.$$

This is an immediate consequence of Theorem 4.1. In fact, (12) is obtained from (8) by setting $x = a$ and applying (6).

(b) Express the following as single logarithms (with coefficient 1):
(i) $3 \log_a x$,
(ii) $3 \log_a x + \log_a y$,
(iii) $3 \log_2 7 + 1$,
(iv) $3 \log_5 6 - 2$,
(v) $\log_4 3 + \log_4 6 - \log_4 9$.

We compute:
(i) $3 \log_a x = \log_a x^3$,
(ii) $3 \log_a x + \log_a y = \log_a x^3 + \log_a y$
$$= \log_a x^3 y,$$
(iii) $3 \log_2 7 + 1 = \log_2 7^3 + \log_2 2$
$$= \log_2 (7^3 \cdot 2)$$
$$= \log_2 686,$$
(iv) $3 \log_5 6 - 2 = \log_5 6^3 - 2 \log_5 5$
$$= \log_5 6^3 - \log_5 5^2$$
$$= \log_5 (6^3/5^2)$$
$$= \log_5 (216/25),$$
(v) $\log_4 3 + \log_4 6 - \log_4 9 = \log_4 \dfrac{3 \cdot 6}{9}$
$$= \log_4 2.$$

(c) Express the following as single logarithms (with coefficient 1):
(i) $3 \log_a x + 2 \log_a y$,
(ii) $\frac{2}{3} \log_{10} 4 + \frac{1}{3} \log_{10} 54 - \frac{2}{3} \log_{10} 2$,
(iii) $2 \log_{10} 2 + 3 \log_{10} 3 - 3$.

We compute:
(i) $3 \log_a x + 2 \log_a y = \log_a x^3 + \log_a y^2$
$$= \log_a x^3 y^2,$$
(ii) $\frac{2}{3} \log_{10} 4 + \frac{1}{3} \log_{10} 54 - \frac{2}{3} \log_{10} 2 = \log_{10} 4^{2/3} + \log_{10} 54^{1/3} - \log_{10} 2^{2/3}$
$$= \log_{10} \left(\frac{4^{2/3} 54^{1/3}}{2^{2/3}} \right)$$

$$= \log_{10}\left(\frac{2^{4/3}2^{1/3}27^{1/3}}{2^{2/3}}\right)$$
$$= \log_{10}(2 \cdot 3)$$
$$= \log_{10} 6,$$

(*iii*) $\quad 2\log_{10} 2 + 3\log_{10} 3 - 3 = \log_{10} 2^2 + \log_{10} 3^3 - \log_{10} 10^3$

$$= \log_{10}\left(\frac{2^2 3^3}{10^3}\right)$$
$$= \log_{10} 0.108.$$

Apart from its intrinsic interest, Theorem 4.1 is of great importance from the computational point of view. Identities (3), (4), and (5) tell us that the logarithm of a product (quotient, power) of positive real numbers is the sum (difference, multiple) of their logarithms. Now in practice, addition, subtraction, and multiplication are easier to perform than multiplication, division, and exponentiation, respectively. Therefore if we have a table of logarithms to a convenient base, we can perform any multiplication, say, by adding the logarithms of the numbers, and then finding the number in the table whose logarithm is equal to this sum. For example, suppose that we want to compute $z = xy$, where

(13) $\qquad\qquad\qquad x = 3.14, \quad y = 1.41.$

From the tables at the end of the book we can find, to four decimal places, that

$$\log_{10} 3.14 = 0.4969,$$
$$\log_{10} 1.41 = 0.1492.$$

(The method of using the tables will be discussed in the next section.) We now compute

$$\log_{10} z = \log_{10} x + \log_{10} y$$
$$= 0.4969 + 0.1492$$
$$= 0.6461,$$

and again, from the tables, we find that

$$\log_{10} 4.43 = 0.6464.$$

Hence, we have found an approximate value for z,

$$z = 3.14 \times 1.41$$
$$= 4.43.$$

Similarly, if we want to compute the quotient $3.14 \div 1.41$, we apply (4) and look for the number whose logarithm is

$$\log_{10} 3.14 - \log_{10} 1.41 = 0.4969 - 0.1492$$
$$= 0.3477.$$

By inspection of the tables we find that

$$\log_{10} 2.23 = 0.3483.$$

Thus

$$3.14 \div 1.41 = 2.23,$$

approximately.

Lastly, suppose that we are more ambitious: We want to evaluate the number $3^{1.414}$, a task that we abandoned at the beginning of this section. Here we evaluate the logarithm of this number:

$$\log_{10} 3^{1.414} = 1.414 \log_{10} 3.$$

From the tables we find that

$$\log_{10} 3 = 0.4771$$

and compute either directly, or by the use of logarithms, that

$$1.414 \times 0.4771 = 0.6746.$$

We have obtained the answer to our problem, i.e., the value of $\log_{10} 3^{1.414}$. It remains to look in the tables and find a number whose logarithm to the base 10 is 0.6746. As a matter of fact,

$$0.6749 = \log_{10} 4.73,$$

and thus

$$3^{1.414} = 4.73,$$

approximately.

The purpose of the preceding example was to demonstrate a method for performing difficult or even apparently impossible computations by means of logarithms. We have not discussed the method of compiling the table of logarithms. This topic is beyond the scope of this book, and we must assume that this rather tedious task has been competently done by some professional mathematician. We shall conclude this section by discussing in detail the application of logarithms to arithmetical calculations, and we shall evolve an efficient computational procedure.

In some of the previous examples we used logarithms to the base 10. In principle, any positive real number different from 1 could be used for the base. As a matter of fact, the base that is most often used in mathematical analysis is an irrational number denoted by e and equal to 2.7182818 . . . . . Logarithms to base e are called *natural logarithms*, or *Napierian logarithms*, after the Scottish mathematician John Napier (1550–1617) who invented logarithms. We shall not even attempt to explain here why logarithms to such a peculiar base are of importance, and we shall certainly not use natural logarithms for computational purposes.

In other parts of mathematics logarithms to the base 2 are occasionally used, and other bases may also be used for special purposes. For arithmetical calculations, however, logarithms to base 10 possess considerable advantages over logarithms to any other base. Suppose, for example, that we have found by one way or another that $\log_{10} 3.45 = 0.5378$ to four decimal places, and suppose that we also require logarithms to the base 10 of 345, 34.5, 0.345, and 0.0345. We proceed as follows:

$$\begin{aligned}
\log_{10} 345 &= \log_{10} (100 \times 3.45) \\
&= \log_{10} 100 + \log_{10} 3.45 \\
&= 2 + 0.5378,
\end{aligned}$$

$$\begin{aligned}
\log_{10} 34.5 &= \log_{10} (10 \times 3.45) \\
&= \log_{10} 10 + \log_{10} 3.45 \\
&= 1 + 0.5378,
\end{aligned}$$

$$\begin{aligned}
\log_{10} 0.345 &= \log_{10} (10^{-1} \times 3.45) \\
&= \log_{10} 10^{-1} + \log_{10} 3.45 \\
&= -1 + 0.5378,
\end{aligned}$$

$$\begin{aligned}
\log_{10} 0.0345 &= \log_{10} (10^{-2} \times 3.45) \\
&= \log_{10} 10^{-2} + \log_{10} 3.45 \\
&= -2 + 0.5378.
\end{aligned}$$

Thus the logarithms to the base 10 of 3.45, 345, 34.5, 0.345 and 0.0345 differ (not unexpectedly) only by an integer. We know, for example, that

$$\log_{10} 100 < \log_{10} 345 < \log_{10} 1000,$$

i.e.,

$$2 < \log_{10} 345 < 3,$$

and that $\log_{10} 3.45 = 0.5378$. Hence $\log_{10} 345$ must be 2.5378. This is true in general. If we know the "fractional part" of a logarithm, we can easily compute the "integer part" and thus the logarithm itself.

It is clear that if we use the ordinary decimal system of notation, logarithms to base 10 offer considerable simplification in the computation and construction of tables. This simplification would not be available if logarithms to other bases were used. For example, if we use base 2 we have $\log_2 3.45 = 1.7867$ while $\log_2 34.5 = 5.1089$.

Definition 4.3 (Common Logarithms)   Logarithms to the base 10 are called *common logarithms.* We shall denote the common logarithm of x simply by $\log x$. It is convenient to split $\log x$ into its integer part and fractional part,

$$\log x = c(x) + m(x),$$

where $c(x)$ is an integer and $0 \leq m(x) < 1$. Clearly $c(x)$ and $m(x)$ are uniquely determined by $\log x$. The number $c(x)$ is called the *characteristic* of $\log x$, and the number $m(x)$ the *mantissa* of $\log x$. In the context of common logarithms, particularly in computations, the number 10^y, where y is any real number, is called the *antilogarithm* of y and is denoted by antilog y. In other words, antilog $= \log^{-1} = \exp_{10}$.

Example 4.3   Given that
$$\log 300 = 2.4771,$$

what are the characteristic and the mantissa of $\log 300$? of $\log 0.3$?

By definition, the characteristic of $\log 300$ is 2 and its mantissa is 0.4771. That is, $\log 3 = 0.4771$. Thus

$$\begin{aligned} \log 0.3 &= \log(10^{-1} \times 3) \\ &= -1 + 0.4771. \end{aligned}$$

Therefore the characteristic of 0.3 is -1 and its mantissa is 0.4771.

If $\log x$ is nonnegative and is written in decimal notation, it is easy to read off its characteristic and its mantissa: The characteristic is the integer part of $\log x$ and the mantissa is its fractional part. If $\log x$ is negative, in a sense it is no longer easy. For example, we have just found that

$$\log 0.3 = -0.5229;$$

but, of course, the characteristic of $\log 0.3$ is not 0, nor is its mantissa 0.5229. We can compute from the above definitions that the characteristic and mantissa are -1 and 0.4771, respectively. As we shall see, it is important to know the characteristic and the mantissa of a logarithm. We therefore introduce the following notation:

$$\begin{aligned} \bar{1}.4771 &= -1 + 0.4771, \\ \bar{2}.4771 &= -2 + 0.4771, \\ \bar{3}.4771 &= -3 + 0.4771, \text{ etc.} \end{aligned}$$

In other words, the bar over the integer part of a logarithm indicates that this integer is preceded by a minus sign.

Before we explain in detail the technique of calculation by means of logarithms, we put together in the form of a theorem some propositions about common logarithms, most of them previously proved for logarithms to any base.

Theorem 4.2 (a) *If x and y are positive real numbers and r any real number, then:*

(14) $$\log (xy) = \log x + \log y,$$

(15) $$\log (x/y) = \log x - \log y,$$

(16) $$\log x^r = r \log x,$$

(17) $$\log 10 = 1,$$

(18) $$\log 1 = 0,$$

(19) $$\text{antilog} (\log x) = x.$$

(b) *If $10^c \leq x < 10^{c+1}$ for some integer c, then*

$$c \leq \log x < c + 1;$$

thus c is the characteristic of log x.
(c) *If c is the characteristic of* log x, *then the mantissa of* log x *is*

$$\log x - c = \log (10^{-c} x).$$

(d) *If k is any integer, then the mantissa of* log $(10^k x)$ *does not depend on k; i.e., the mantissa of* log $(10^k x)$ *for any integer k is equal to the mantissa of* log x.

SYNOPSIS

In this section we explained the meaning of powers with irrational exponents, and we defined $\exp_a$, the exponential function to base a. We then defined the logarithmic function to base a as the inverse function of $\exp_a$. We showed that to any base $a > 0$, $a \neq 1$, and for any positive real numbers x and y,

$$\log_a (xy) = \log_a x + \log_a y,$$
$$\log_a (x/y) = \log_a x - \log_a y,$$

and

$$\log_a (x^r) = r \log_a x,$$

for any real number r.

QUIZ

Answer *true* or *false:*

1. $3^{1/3} > 4^{1/4}$.
2. $2^{1/2} \cdot 2^{1/3} = 2^{1/6}$.
3. If a and b are positive real numbers, then $(a/b)^n = a^n/b^n$ for any real number n.
4. $2^{1/n} = 1/2^n$.
5. $10^{1+\log x} = 10x$.
6. $-\log 5 = \log(-5)$.
7. $\log(\log 10) = 0$.
8. $(\sqrt{10})^{\log 3} = \sqrt{3}$.
9. $\log_9 25 = \log_3 5$.
10. $\log 5 = 1 - \log 2$.

EXERCISES

1. Find the values of:

   (a) $9^{1/2}$, (b) $1^{1/2}$, (c) 1^0,
   (d) 2^0, (e) 2^{-1}, (f) 1^{-2},
   (g) $4^{-1/2}$, (h) 4^{-2}, (i) $(1/5)^{-1}$,
   (j) 0.1^{-2}, (k) $8^{1/3}$, (l) $16^{1/4}$,
   (m) $16^{-1/4}$, (n) $(\frac{1}{4})^{1/2}$, (o) $(\frac{1}{4})^0$,
   (p) $0.04^{-1/2}$, (q) $(-0.5)^{-2}$, (r) $(-1)^{-1}$,
   (s) $(3/2)^{-2}$, (t) $0.001^{-1/3}$, (u) 0.1^{-2},
   (v) $32^{1/5}$.

   (Express the results as integers or as fractions.)

2. Find the values of:

   (a) $4^{3/2}$, (b) $0^{3/2}$, (c) $(1/9)^{1/2}$,
   (d) $125^{2/3}$, (e) $4^{-3/2}$, (f) $16^{3/4}$,
   (g) $64^{2/3}$, (h) $25^{3/2}$, (i) $25^{-1/2}$,
   (j) $8^{-2/3}$, (k) $(\frac{1}{8})^{-2/3}$, (l) $0.09^{3/2}$,
   (m) $0.01^{-3/2}$, (n) $(1/27)^{4/3}$, (o) $(\frac{1}{4})^{-3/2}$,
   (p) $4^{5/2}$, (q) $0.36^{3/2}$, (r) $16^{3/4}$.

   (Express the results as integers or as fractions.)

3. Find the values of x satisfying:

   (a) $2^x = 8$; (b) $8^x = 2$; (c) $8^x = 0.5$;
   (d) $(\frac{1}{2})^x = 8$; (e) $(\frac{1}{8})^x = 8$; (f) $4^x = 8$;
   (g) $16^x = 8$; (h) $8^x = 4$; (i) $8^x = 16$;
   (j) $8^x = \frac{1}{4}$; (k) $8^x = 1$; (l) $8^x = 1/16$;
   (m) $8^x = 8$; (n) $(\frac{1}{8})^x = 1$.

4. Find the values of x satisfying:
   (a) $9^x = 3$;
   (b) $25^x = 5$;
   (c) $25^x = 125$;
   (d) $27^x = \frac{1}{9}$;
   (e) $0.0001^x = 0.1$;
   (f) $(1/9)^x = 1/27$;
   (g) $4^x = 32$;
   (h) $(-5)^x = 0.04$;
   (i) $(0.09)^x = 10/3$;
   (j) $(0.09)^x = 1$;
   (k) $(2/3)^x = 2.25$;
   (l) $(\frac{1}{4})^x = 0.25$;
   (m) $125^x = 0.2$;
   (n) $1.25^x = 1$.

5. Show that if $x \geq 1$, then the characteristic of $\log x$ is one less than the number of digits to the left of the decimal point in x. If $x < 1$ and the first non-zero digit occurs in the kth decimal place of x, show that the characteristic of $\log x$ is $\bar{k}$. (Recall that $\bar{k}$ simply means $-k$.)

6. Solve for x:
   (a) $\log_2 x = 3$;
   (b) $\log_x 1/9 = -2$;
   (c) $\log_x 5 = \frac{1}{2}$;
   (d) $\log_8 1 = x$;
   (e) $\log_8 8 = x$;
   (f) $\log_x 16 = 4$;
   (g) $\log_x 32 = 5/4$;
   (h) $\log_3 x = 3$;
   (i) $\log_5 x = 2$;
   (j) $\log_3 (1/27) = x$;
   (k) $\log_3 (1/81) = x$;
   (l) $\log_3 x = 1$;
   (m) $\log_3 x = 0$;
   (n) $\log_4 x = \frac{1}{2}$;
   (o) $\log_2 x = -3$;
   (p) $\log_x 2 = \frac{1}{2}$;
   (q) $\log_x 64 = 6$;
   (r) $\log_x 2 = -\frac{1}{3}$;
   (s) $\log_9 3 = x$;
   (t) $\log_3 9 = x$;
   (u) $\log_9 x = \frac{1}{2}$.

7. Evaluate:
   (a) $\log 1$,
   (b) $\log 10$,
   (c) $\log 100$,
   (d) $\log 0.1$,
   (e) antilog 0,
   (f) antilog (-1),
   (g) antilog 2,
   (h) antilog (-2),
   (i) antilog $(2 \log 12)$,
   (j) antilog $(\log 2 + \log 3)$,
   (k) antilog $(\log 2 - \log 3)$.

8. Given that $\log 2 = 0.3010$ and $\log 3 = 0.4771$, evaluate the following without the use of tables:
   (a) $\log 6$,
   (b) $\log 5$,
   (c) $\log 18$,
   (d) $\log 1.5$,
   (e) $\log 8/9$,
   (f) $\log 7.5$,
   (g) $\log 0.045$,
   (h) $\log \frac{1}{3}$,
   (i) $\log 1.6$,
   (j) $\log 12.5$.

9. Express each of the following in the form $\log x$:
   (a) $\log 126 - \log 2$,
   (b) $\log 27 + \log 8 - \log 6$,
   (c) $\frac{1}{2} \log 51$,
   (d) $-\log 0.04$,
   (e) $-\frac{1}{3} \log 8$,
   (f) $2 \log 3 - 3 \log 2$,
   (g) $2 - \log 4$,
   (h) $1 + \log 3 - 2 \log 5$,
   (i) $\frac{1}{2} \log 16 - 1$.

2.5 Logarithmic Computations

Theorem 4.2(d) in the preceding section tells us that for the purpose of finding the mantissa of $\log x$, we may disregard the decimal point in the ordinary decimal representation of x, or add as many zeros as we please to the right of it. For example, the mantissas of log 345000, log 3450, log 3.45, log 0.345 and log 0.00345 are all equal to 0.5378. The table of common logarithms in the Appendix (Table 1) lists the logarithms of numbers from 1.00 to 9.99, to four decimal places, and allows us to find the mantissa of the common logarithm of any number, to four decimal places.

The expression "log $x = y$ to four decimal places" (or "correct to four decimal places") means that $-0.00005 \leq y - \log x \leq 0.00005$, i.e., that log x differs from y by not more than 0.00005. Our tables give logarithms correct to four decimal places. To avoid repetition we shall not specify each time that the logarithms are "correct to four decimal places." Thus if we write

$$\log 4.5 = 0.6532,$$

we shall mean

$$0.65315 \leq \log 4.5 \leq 0.65325.$$

In the computations that follow, each logarithm carries a potential error of 0.00005. Thus the logarithm obtained as a result of several operations (addition, subtraction, etc.) may carry a substantially greater error. However, we shall not attempt to estimate this error, and we shall still use, somewhat loosely, the sign of equality. For instance, log 3.5 = 0.5441 and we may write, for example,

$$\begin{aligned} \log (3.5^8) &= 8 \log 3.5 \\ &= 8 \times 0.5441 \\ &= 4.3528, \end{aligned}$$

although we can actually compute directly that $3.5^8 = 22{,}518.75390625$ and check that to four decimal places, the correct value of log (3.5^8) is 4.3522. In other words, in our computations we use the symbol for equality, $=$, to denote that the equated numbers are approximately equal; the degree of approximation will vary from example to example, but all initial values will be read from our table and will be correct to four decimal places.

The method of using Table 1 is very simple and is best illustrated by examples.

Example 5.1 (a) Find the mantissas of log 352 and log 820.

In the first column of Table 1 find the number 3.5. The required mantissa can be found in the same row, in the column headed by 2. We read off .5465. Therefore the mantissa of log 352 is 0.5465. Similarly, we find the mantissa of log 820 in the row "8.2" and in the column "0": 0.9138.

(b) Find the mantissas of log 31, log 2, log 0.3.

We know from Theorem 4.2(d) that these mantissas are equal to the mantissas of log 3.10, log 2.00, log 3.00, respectively. We find these easily: 0.4914, 0.3010, 0.4771.

(c)   Find the mantissas of log 24.762 and of log 0.1472, rounding off the numbers to three significant figures.

We require the mantissas of log 24.8 and of log 0.147. These are equal to the mantissas of log 2.48 and log 1.47, and we find from the table that the mantissas are 0.3945 and 0.1673, respectively.

We do not require any fancy methods to find the characteristic of log x; Theorem 4.2(b) shows us how to find it without difficulty. For example, if we want the characteristic of log 147, we note that

$$2 = \log 100 < \log 147 < \log 1000 = 3,$$

and therefore the characteristic is 2.

Example 5.2   (a)   Evaluate log 147.

We have just found that the characteristic of log 147 is 2. In Example 5.1(c) we found that the mantissa of 147 is 0.1673. Therefore, to four decimal places,

$$\log 147 = 2.1673.$$

(b)   Find log 0.0147.

Since

$$-2 = \log 10^{-2} = \log 0.01 < \log 0.0147 < \log 0.1 = \log 10^{-1} = -1,$$

we conclude that the characteristic of log 0.0147 is $\bar{2}$. Now, the mantissa of log 0.0147 is the same as that of log 147. Hence

$$\log 0.0147 = \bar{2}.1673.$$

(c)   Evaluate log 1472.

In an analogous way we find that

$$\log 1472 = 3.1673.$$

Note that although we look in Table 1 for the mantissa of log 147, we do not compute the characteristic of log 147 but that of log 1472 (or of log 1470).

Table 1 can be used to find antilogarithms as well as logarithms. For example, suppose that we want to find the value of antilog 2.4382. We are looking for a number y such that $\log y$ has characteristic 2 and mantissa 0.4382. We inspect the table to locate the number whose log is 0.4382. We find that $\log 2.74 = 0.4378$ and $\log 2.75 = 0.4393$. Since the first of these differs from 0.4382 by less than the second, we write the information to three significant figures: antilog $0.4382 = 2.74$. Now, the characteristic of 2.4382 is 2. Thus antilog 2.4382 is a number between 100 and 1000, and we conclude that antilog $2.4382 = 274$.

Example 5.3 Evaluate antilog $\bar{2}.6347$, correct to three significant figures.
We find by inspection that the number appearing in the table closest to 0.6347 is 0.6345 and that antilog $0.6345 = 4.31$. Now,

$$0.01 \leq \text{antilog } \bar{2}.6347 < 0.1,$$

since the characteristic of the logarithm of antilog $\bar{2}.6347$ is $\bar{2}$. It follows that

$$\text{antilog } \bar{2}.6347 = 0.0431.$$

Finally, we demonstrate in several examples the details of arithmetical calculations by means of logarithms.

Example 5.4 Evaluate
(a) $x = 5270 \times 24.8$,
(b) $x = 52.7 \times 0.0248$,
(c) $x = 5270 \div 24.8$,
(d) $x = 52.7 \div 0.0248$.

(a)

Numbers	Logs	Antilogs
5270	3.7218	
$\times$ 24.8	$+$ 1.3945	
	5.1163	$131,000 = x$

(b)

Numbers	Logs	Antilogs
52.7	1.7218	
$\times$ 0.0248	$+ \bar{2}.3945$	
	0.1163	$1.31 = x$

Note the method used in the "Logs" column to compute the sum of 1.7218 and $\bar{2}.3945$: After the mantissas are added, 1 is "carried forward." The characteristic of the sum is therefore the sum of two characteristics, 1 and $\bar{2}$, and of the number 1 carried forward: $1 + \bar{2} + 1 = 0$.

(c)

Numbers	Logs	Antilogs
5270	3.7218	
÷ 24.8	− 1.3945	
	2.3273	212 = x

(d)

Numbers	Logs	Antilogs
52.7	1.7218	
÷ 0.0248	− $\bar{2}$.3945	
	3.3273	2120 = x

Here the second characteristic is negative and we subtract it from the characteristic of log 52.7 as follows:

$$1 - \bar{2} = 1 - (-2)$$
$$= 3.$$

Example 5.5 Evaluate

(a) $x = (65.7)^3$,
(b) $x = \sqrt[3]{65.7}$,
(c) $x = (0.657)^3$,
(d) $x = \sqrt[3]{0.657}$.

(a)

Numbers	Logs	Antilogs
65.7	1.8176	
65.7^3	3 × 1.8176	
	= 5.4528	284,000 = x

(b)

Numbers	Logs	Antilogs
65.7	1.8176	
$\sqrt[3]{65.7}$	$\frac{1}{3}$ × 1.8176	
	= 0.6059	4.04 = x

It cannot be stressed too much that each computation with logs should be checked by a rough computation with the actual numbers in case the decimal point has been mistakenly put in the wrong place. Thus in Example 5.5(a), the student should mentally compute that $(65.7)^2$ is just over 4000, and therefore $(65.7)^3$ is of the same order of magnitude as 250,000. In Example 5.5(b) we mentally check that $(4.04)^3$ is slightly more than 64, and thus our result is of the right order of magnitude. If, by mistake, we had obtained $x = 0.404$, the error would be noticed at once, since $(0.404)^3$ cannot possibly be 65.7.

(c)

Numbers	Logs	Antilogs
0.657	$\bar{1}.8176$	
0.657^3	$3 \times \bar{1}.8176$	
	$= \bar{1}.4528$	$0.284 = x$

Note that

$$3 \times \bar{1}.8176 = 3(-1 + 0.8176)$$
$$= -3 + 2.4528$$
$$= -1 + 0.4528$$
$$= \bar{1}.4528.$$

(d)

Numbers	Logs	Antilogs
0.657	$\bar{1}.8176$	
$\sqrt[3]{0.657}$	$\frac{1}{3} \times \bar{1}.8176$	
	$= \bar{1}.9392$	$0.869 = x$

Here the division of $\bar{1}.8176$ by 3 is a bit tricky. It is done as follows:

$$\tfrac{1}{3} \times \bar{1}.8176 = \tfrac{1}{3}(-1 + 0.8176)$$
$$= \tfrac{1}{3}(-3 + 2.8176)$$
$$= -1 + 0.9392$$
$$= \bar{1}.9392.$$

If we had to compute $\tfrac{1}{3} \times \bar{2}.8176$, we would proceed thus:

$$\tfrac{1}{3} \times \bar{2}.8176 = \tfrac{1}{3}(-2 + 0.8176)$$
$$= \tfrac{1}{3}(-3 + 1.8176)$$
$$= -1 + 0.6059$$
$$= \bar{1}.6059.$$

In the next example we show how to perform more complicated calculations of the kind often encountered in evaluating empirical formulae in physics or other sciences.

Example 5.6 (a) The average energy E of a vibrating particle is given in physics by the formula

$$E = \frac{m\pi^2 a^2}{T^2}.$$

Compute E if $m = 0.065$, $a = 0.235$, $T = 0.00285$, $\pi = 3.14$.

Numbers	Logs		Antilogs
0.065	$\bar{2}.8129$	$\bar{2}.8129$	
$\times 3.14^2$	2×0.4969		
	$= 0.9938$	$+ 0.9938$	
$\times 0.235^2$	$2 \times \bar{1}.3711$		
	$= \bar{2}.7422$	$+ \bar{2}.7422$	
		$\bar{2}.5489$	
$\div 0.00285^2$	$2 \times \bar{3}.4548$		
	$= \bar{6}.9096$	$- \bar{6}.9096$	
		3.6393	$4360 = E$

The only detail of calculation which may cause some difficulty is the subtraction of $\bar{6}.9096$ from $\bar{2}.5489$. This is computed as follows:

$$\bar{2}.5489 - \bar{6}.9096 = -3 + 1.5489 - (-6 + 0.9096)$$
$$= 3 + 0.6393$$
$$= 3.6393.$$

(b) The acceleration of a falling body in a resisting medium is given in mechanics by the formula
$$a = ge^{-kt/m}.$$

Find the value of a if $g = 981$, $e = 2.72$, $k = 22.5$, $t = 2$, $m = 50$.
   We have
$$a = 981 \times 2.72^{-22.5 \times 2/50}$$
$$= 981 \times 2.72^{-0.9},$$

and we calculate a by means of logarithms.

Numbers	Logs		Antilogs
981	2.9917	2.9917	
$\times 2.72^{-0.9}$	-0.9×0.4346		
	$= -0.39114$	-0.3911	
		2.6006	$399 = a$

(c) If P is invested at j percent for n years, the compound amount S, is given by the formula
$$S = P(1 + j/100)^n.$$

Find S if $P = 785$, $j = 6$, $n = 12$.

We first compute $1 + j/100$, since this cannot be done by means of logarithms. It is easy, however, to calculate that $1 + j/100 = 1.06$ and continue our calculation using logarithms.

Numbers	Logs		Antilogs
785	2.8949	2.8949	
$\times 1.06^{12}$	12×0.0253		
	$= 0.3036$	$+ 0.3036$	
		3.1985	$1580 = S$

In Example 5.2(c) we were asked to evaluate log 1472. We found that log 1470 = 3.1673, and we concluded that log 1472 is approximately 3.1673. Indeed, this is the best approximation that can be read off the tables, since log 1472 is closer to log 1470 than to log 1480 = 3.1703. We know, of course, that

$$3.1673 < \log 1472 < 3.1703,$$

i.e., that the actual value of log 1472 lies somewhere between 3.1673 and 3.1703. If we assume that for a small change in x, the change in log x is proportional to the change in x, we can conclude that

$$\frac{\log 1472 - \log 1470}{\log 1480 - \log 1470} = \frac{1472 - 1470}{1480 - 1470},$$

i.e.,

$$\log 1472 = \log 1470 + \tfrac{2}{10} (\log 1480 - \log 1470)$$
$$= 3.1673 + 0.2 \times 0.0030$$
$$= 3.1679.$$

This approximating procedure is called *linear interpolation* (or simply *interpolation*), since in our computation we have essentially assumed that the portion of the graph $y = \log x$ between the points whose x-coordinates are 1470 and 1480 approximates a straight line. The method of interpolation is sufficiently accurate to allow us to extend the use of tables to numbers given to four-figure accuracy and evaluate the antilogs to the same number of significant figures. The following examples illustrate the method in detail.

Example 5.7 Evaluate to four significant figures:

(a) log 23.94,

(b) antilog $\overline{2}.7569$,

(c) 3.576^3,

(d) $\sqrt{0.06344} \times \sqrt[3]{16.27}$.

(a)   We find from Table 1 that

$$\log 23.90 = 1.3784 \quad \text{and} \quad \log 24.00 = 1.3802.$$

The number 23.94 is $\frac{4}{10} = 0.4$ of the way from 23.90 to 24.00. The method of linear interpolation assumes that $\log 23.94$ is 0.4 of the way from $\log 23.90$ to $\log 24.00$. Since the difference between 1.3802 and 1.3784 is 0.0018, we have

$$\begin{aligned}
\log 23.94 &= 1.3784 + 0.4 \times 0.0018 \\
&= 1.3784 + 0.0007 \\
&= 1.3791.
\end{aligned}$$

(b)   From the table we obtain

$$\text{antilog } \bar{2}.7566 = 0.05710 \quad \text{and} \quad \text{antilog } \bar{2}.7574 = 0.05720,$$

and thus

$$\begin{aligned}
\text{antilog } \bar{2}.7569 &= 0.05710 + \frac{\bar{2}.7569 - \bar{2}.7566}{\bar{2}.7574 - \bar{2}.7566}(0.05720 - 0.05710) \\
&= 0.05710 + \frac{3}{8} \times 0.00010 \\
&= 0.05714.
\end{aligned}$$

(c)   We compute:

$$\begin{aligned}
\log 3.576 &= \log 3.570 + 0.6 (\log 3.580 - \log 3.570) \\
&= 0.5527 + 0.6 \times 0.0012 \\
&= 0.5534.
\end{aligned}$$

Numbers	Logs	Antilogs
3.576	0.5534	
3.576^3	3×0.5534	
	$= 1.6602$	antilog 1.6602

Thus

$$\begin{aligned}
3.576^3 &= \text{antilog } 1.6602 \\
&= 45.70 + 0.3 \times 0.1 \\
&= 45.73.
\end{aligned}$$

(d)   We first evaluate $\log 0.06344$ and $\log 16.27$:

$$\log 0.06344 = \log 0.06340 + 0.4(\log 0.06350 - \log 0.06340)$$
$$= \bar{2}.8021 + 0.4 \times 0.0007$$
$$= \bar{2}.8024;$$

$$\log 16.27 = \log 16.20 + 0.7(\log 16.30 - \log 16.20)$$
$$= 1.2095 + 0.7 \times 0.0027$$
$$= 1.2114.$$

Numbers	Logs		Antilogs
$\sqrt{0.06344}$	$\frac{1}{2} \times \bar{2}.8024$ $= \bar{1}.4012$	$\bar{1}.4012$	
$\times \sqrt[3]{16.27}$	$\frac{1}{3} \times 1.2114$ $= 0.4038$	$+ 0.4038$	
		$\overline{1}.8050$	antilog $\bar{1}.8050$ $= 0.6383$

For antilog $\bar{1}.8050 = 0.6380 + \frac{2}{7} \times 0.0010 = 0.6383.$

SYNOPSIS

In this section we explained how to use the table of common logarithms (Table 1 in the Appendix), and we developed an efficient method for numerical calculations by means of logarithms. We also showed how to use linear interpolation to find the logarithm of a number with four significant figures.

QUIZ

Answer *true* or *false:*
1. log 0.35 and log 305 have equal mantissas.
2. $\log 0.35 = 0.5441.$
3. $\log 0.001 = \bar{3}.0000.$
4. $\log 100 = 3.0000.$
5. $\log 1 = 0.0000.$
6. $\log (-1) = -0.0000 = 0.0000.$
7. $\log \sqrt{0.1} = \frac{1}{2} \times \bar{1}.0000.$
8. $\frac{1}{2} \times \bar{1}.0000 = \bar{1}.5000.$
9. $\log 5^{-1} = (\log 2) - 1.$
10. $\log 45670 - \log 0.04567$ is an integer.

EXERCISES

1. Find the common logarithms of the following numbers:

   (a) 1.23,   (b) 0.0123,   (c) 123,
   (d) 12,300,   (e) 0.02,   (f) 20,000,
   (g) 0.0003,   (h) 0.0002,   (i) 0.3,
   (j) 0.30,   (k) 300,000,   (l) 801,
   (m) 8.1,   (n) 1.02,   (o) 6,000,
   (p) 6,010,   (q) 0.061,   (r) 6.01,
   (s) 69.3,   (t) 0.00305.

2. Find the antilogarithms of the following numbers:

   (a) 1.8451,   (b) 3.8451,   (c) $\bar{2}$.8451,
   (d) $\bar{1}$.8500,   (e) 0.1399,   (f) 1.3139,
   (g) 0.1673,   (h) 1.6730,   (i) 1.1673,
   (j) 4.9786,   (k) $\bar{3}$.6990,   (l) 0.9410,
   (m) 5.6776,   (n) 2.6021,   (o) 1.6464,
   (p) 1.6474,   (q) 1.6484,   (r) $\bar{5}$.5551.

3. Find the antilogarithms of the following numbers correct to four significant figures (i.e., take the antilogs of the mantissas correct to third decimal place).

   (a) 0.5720,   (b) 3.570,   (c) $\bar{2}$.570,
   (d) 0.9328,   (e) 4.9328,   (f) 1.9328,
   (g) 0.4139,   (h) $\bar{3}$.4139,   (i) 1.7074,
   (j) $\bar{1}$.0400,   (k) 5.7780,   (l) 5.8803,
   (m) 4.9317,   (n) 4.9318,
   (o) $\bar{3}$.6 [*Hint:* antilog $\bar{3}$.6 = antilog $\bar{3}$.6000.],   (p) 2.51,
   (q) 0.4,   (r) 4.14,   (s) 1.235.

4. Perform the following computations by means of logarithms:

   (a) 3.45×2.93,   (b) 0.123×0.132,
   (c) 0.874×0.087,   (d) 543×0.0231,
   (e) 15400×0.34,   (f) $(0.537)^3$,
   (g) $345/24.6$,   (h) $4.56/2.97$,
   (i) $0.34/0.0567$,   (j) $52/0.973$,
   (k) $1/0.0425$,   (l) $123/45700$.

5. Perform the following computations by means of logarithms:

   (a) $\dfrac{35.6 \times 526}{345}$,   (b) $\dfrac{0.487 \times 13.5}{2.37}$,

   (c) $\dfrac{35.6 \times 0.067}{391}$,   (d) $\dfrac{0.235 \times 0.0173}{0.0827}$,

   (e) $\dfrac{47.6}{8530 \times 2470}$,   (f) $\dfrac{0.0232}{4.83 \times 0.193}$,

   (g) $\dfrac{26.9 \times 3.45}{0.611 \times 459}$,   (h) $\dfrac{0.534 \times 28}{6300 \times 471}$,

(i) $\dfrac{0.073 \times 24.8}{5432 \times 0.069}$,

(j) $\dfrac{5478 \times 0.2351}{0.043 \times 2.387}$,

(k) $\dfrac{4.766 \times 0.0351}{2950 \times 30.78}$,

(l) $\dfrac{2370 \times 123}{0.024 \times 0.112}$,

(m) $\dfrac{0.568 \times 23.81}{0.039 \times 0.00273}$,

(n) $\dfrac{0.639^2}{0.456 \times 47.8}$,

(o) $(5.06)^8$,

(p) $\sqrt{5.06}$,

(q) $64.5^4 \times 0.231^5$,

(r) $\sqrt[3]{0.29} \times 3.05$,

(s) $\dfrac{\sqrt{472}}{0.856}$,

(t) $\sqrt{0.56}$,

(u) $\sqrt[3]{0.56}$,

(v) $\sqrt[3]{0.056}$,

(w) $\sqrt[3]{0.0056}$.

6. Perform the following computations by means of logarithms:

(a) $\dfrac{25.7\sqrt{385}}{0.467}$,

(b) $37.3\sqrt[4]{\dfrac{26.3}{47.5}}$,

(c) $\sqrt{\dfrac{96.3}{458 \times 0.732}}$,

(d) $\left(\dfrac{6450}{16.7\sqrt[3]{0.236}}\right)^{2/3}$,

(e) $\dfrac{278^{3/2}}{\sqrt{47.8}}$,

(f) $\dfrac{27.6 \times 345^{1/2}}{6.75^{2/3}}$,

(g) $3.14\sqrt{\dfrac{38 \times 415}{675}}$,

(h) $37.5\left(\dfrac{571}{493}\right)^{1.41}$,

(i) $0.0023^{2/5}$,

(j) $\dfrac{3.14\sqrt{576}}{4.21\sqrt{499}}$,

(k) $\sqrt[3]{\dfrac{232}{673}}$,

(l) 0.853^{-3}.

7. Solve the following equations for x:
   (a) $8^x = 31$;
   (b) $8^x = 10$;
   (c) $0.8^x = 10$;
   (d) $5^{2x} = 15$;
   (e) $14.93^x = 1$;
   (f) $3.75^x = 9.02$;
   (g) $3^x = 2.17 \times 4^x$ [*Hint:* Divide both sides of the equation by 4^x.];
   (h) $(5.72^x - 62)/51 = 1.73$;
   (i) $0.59^x = 68 \times 2.13^x$;
   (j) $521 \times 4.7^x = 291 \times 9.13^x$.

8. What is the volume of a room whose dimensions in feet are 26.8 by 11.9 by 10.8?

9. The period of a simple pendulum of length l is given by the formula $t = \pi\dfrac{l}{g}$, where $g = 981$ and $\pi = 3.14$.
   (a) Find t if $l = 37.8$.
   (b) Find l if $t = 1.35$.

Trigonometric Functions

3.1 Trigonometric Functions of Acute Angles

Although many propositions in Euclid's *Elements* express relations between the sides and angles of a triangle, few of these relations are metric in nature. That is to say, Euclid's theorems often concern inequalities between angles and sides of a polygon (e.g., "the largest angle in a triangle lies opposite the longest side."), but do not specify the numerical relations between the angles and the sides. *Trigonometry* is precisely that part of mathematics that attempts to relate the lengths of the sides of a polygon to its angles.

In the present section we introduce trigonometric functions of acute angles only. A general discussion is postponed to subsequent sections of this chapter. We assume that the reader is acquainted with the basic concepts of elementary geometry: angles, equality of angle measure, congruence (i.e., "equality") of triangles, properties of right-angled triangles, isosceles triangles, etc.

The basic unit of measurement of angles is the right angle. For convenience the right angle is divided into 90 equal parts, each of which is said to have the measure of one *degree*, written 1°. Thus a right angle is equal to 90°, one half of a right angle to 45°, one third of a right angle to 30°, etc. An angle of 1° is further subdivided into 60 equal parts, each of which is said to have the measure of one *minute*, written 1′. Thus

$$1° = 60′,$$
$$(\tfrac{1}{2})° = 30′,$$
$$(\tfrac{1}{4} \text{ of a right angle}) = 22.5° = 22°30′,$$
$$(\tfrac{1}{8} \text{ of a right angle}) = 11°15′,$$
$$(\tfrac{1}{100} \text{ of a right angle}) = 54′.$$

Let *ABC* be a triangle.

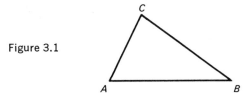

Figure 3.1

We denote the measure of the angle at vertex A by $\angle BAC$, etc. Whenever possible we may also use the following convenient, if somewhat ambiguous, notation: We denote the measure of any angle in the triangle ABC by the same letter as the vertex. Thus $\angle BAC$ is denoted by A, $\angle ACB$ by C, $\angle CBA$ by B, and the well-known theorem on the sum of angles in a triangle can be stated:

(1) $$A + B + C = 180°.$$

The lengths of the sides AB, BC, CA will be denoted either simply by AB, BC, CA or, if possible, by small letters c, a, b (opposite vertices C, A, B), respectively.

Recall that two triangles are *similar* if they are equiangular. That is to say, if in triangles ABC and PQR (Figure 3.2)

Figure 3.2

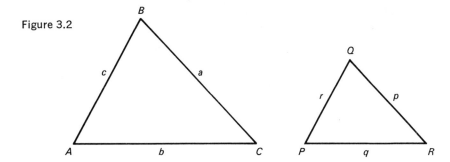

we have

$$A = P, \quad B = Q, \quad C = R,$$

then

$$a/p = b/q = c/r,$$

or equivalently,

(2) $$a/b = p/q, \quad b/c = q/r, \quad c/a = r/p,$$

where p, q, r denote the lengths of sides QR, PR, PQ, respectively.

Observe that if $A = P$ and $C = R$, then by (1), $B = Q$ and the two triangles are similar. In particular, if the triangles happen to be right-angled (at C and R, say) then if $A = P$ (or equivalently if $B = Q$), then the triangles are equiangular, and therefore similar. This observation about right-angled triangles has some remarkable consequences.

Let ABC and PQR be two triangles right-angled at C and R, respectively, and suppose that the angles A and P are equal; $A = P = \alpha$, say. Then the triangles ABC and PQR are similar, and by (2) we have

$$a/b = p/q, \qquad a/c = p/r, \qquad b/c = q/r.$$

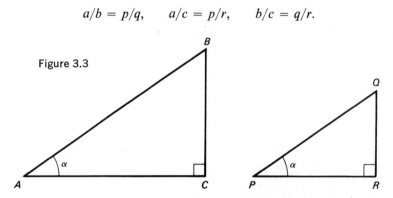

Figure 3.3

Now, the only assumptions we have made about the two triangles is that they are right-angled and that one of the acute angles is equal to α. It follows therefore that for any right-angled triangle with one angle equal to α ($0° < \alpha < 90°$), each of the following three ratios has a fixed value (see Figure 3.4):

(3)

$$\frac{\text{length of opposite side}}{\text{length of adjacent side}},$$

$$\frac{\text{length of opposite side}}{\text{length of hypotenuse}},$$

$$\frac{\text{length of adjacent side}}{\text{length of hypotenuse}}.$$

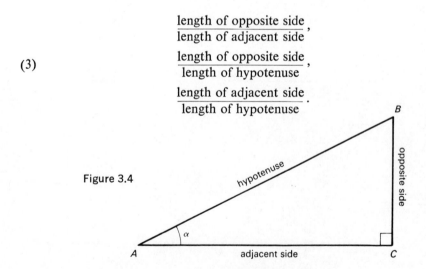

Figure 3.4

In other words, each of these ratios is a function of α alone and does not depend on the particular triangle used to evaluate it.

Definition 1.1   (Trigonometric Functions of Acute Angles)   Let S denote the set of acute angles.   Let ABC be a triangle right-angled at C and let $\angle BAC$ be equal to α.   We define three functions, *sine, cosine, tangent* (abbreviated sin, cos, tan) from S to R, the set of real numbers, as follows: sin: $S \to R$ is defined by

(4) $$\sin \alpha = a/c$$

for all $\alpha \in S$; cos: $S \to R$ is defined by

(5) $$\cos \alpha = b/c$$

for all $\alpha \in S$; tan: $S \to R$ is defined by

(6) $$\tan \alpha = a/b.$$

These functions are called *trigonometric functions.*

Since any right-angled triangle with one of its acute angles equal to α is similar to the triangle ABC, the above functions are well-defined.

It is clear from the geometry of right-angled triangles that the image set of both the sine and the cosine function is $\{x \mid 0 < x < 1\}$, and the image set of the tangent function is the set of positive real numbers.

It cannot be stressed too much that sine, cosine, and tangent are real-valued functions and not geometric properties of triangles nor measurements of the sides of a triangle.   The use of a right-angled triangle in Definition 1.1 is just a device for obtaining the values of these functions at α.

Example 1.1   Determine the values of sin α, cos α, and tan α, where α is the angle A in the triangle ABC, right-angled at C:
(a) $a = 3$ cm., $b = 4$ cm.;
(b) $a = 1$ in., $b = 2$ in.;
(c) $\alpha = 40°$.

Figure 3.5

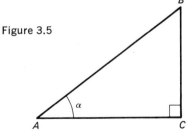

(a) Recall the Pythagorean theorem:

$$c^2 = a^2 + b^2.$$

Thus if $a = 3$ cm. and $b = 4$ cm.,

$$c = \sqrt{3^2 + 4^2}$$
$$= 5 \text{ cm.,}$$

and by Definition 1.1

$$\sin \alpha = 3/5,$$
$$\cos \alpha = 4/5,$$
$$\tan \alpha = 3/4.$$

(b) We compute that

$$c = \sqrt{1^2 + 2^2}$$
$$= \sqrt{5} \text{ in.}$$

Therefore

$$\sin \alpha = \frac{1}{\sqrt{5}},$$
$$\cos \alpha = \frac{2}{\sqrt{5}},$$
$$\tan \alpha = \tfrac{1}{2}.$$

(c) We draw any convenient triangle ABC, right-angled at C and with $\angle BAC = 40°$. Let $b = 2$ in., for example. Construct triangle ABC and measure the lengths a and c. We obtain from Figure 3.6

$$a = 1.7 \text{ in.,}$$
$$c = 2.6 \text{ in.,}$$

approximately. Hence

$$\sin 40° = 1.7/2.6 = 0.65,$$
$$\cos 40° = 2/2.6 = 0.77,$$
$$\tan 40° = 1.7/2 = 0.85,$$

approximately.

Figure 3.6

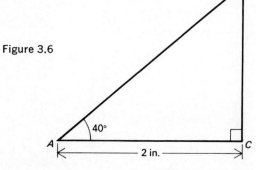

It is obvious from Example 1.1(c) that the value of any of the trigonometric functions for a given angle can be found by measuring the length of a segment on an appropriate diagram.   Such a procedure, of course, is neither efficient nor accurate. Fortunately there are analytical methods, which we cannot discuss here, for computing the values of trigonometric functions to any required degree of accuracy. Using these methods, values of trigonometric functions have been computed and set up in the form of tables similar to the tables of logarithms. In Table 2 of the Appendix, the values of the trigonometric functions are given to 4 places for each angle between 0° and 90° at 30′ intervals. The values of the sine, cosine, and tangent functions can be read off directly from the tables. For example, we find that

$$\sin 40° = 0.6428,$$
$$\cos 40° = 0.7660,$$
$$\tan 40° = 0.8391.$$

In the same manner, if we want to find the sine, the cosine, or the tangent of any other acute angle α, we can find it in the row which has α in the first column (under "Degrees").

We can use linear interpolation to find the value of a trigonometric function for angles between those given in the table. The method is similar to the one used in Section 2.5 to interpolate between values in logarithm tables. We illustrate it in the following examples.

Example 1.2   (a) Evaluate $\sin 25°36′$.
   We find from Table 2 that

$$\sin 25°30′ = 0.4305 \quad \text{and} \quad \sin 26° = 0.4384.$$

We assume again that, for a small change in the angle, the change in the value of the sine function is proportional to the change in the angle. Thus

$$\sin 25°36′ = \sin 25°30′ + \frac{25°36′ - 25°30′}{26° - 25°30′} (\sin 26° - \sin 25°30′)$$

$$= 0.4305 + \frac{6}{30} \times 0.0079$$

$$= 0.4321.$$

(b) Find $\tan 41°20′$.
Again from Table 2:

$$\tan 41° = 0.8693 \quad \text{and} \quad \tan 41°30′ = 0.8847.$$

Therefore

$$\tan 41°20' = 0.8693 + \frac{20}{30}(0.8847 - 0.8693)$$
$$= 0.8693 + 0.0103$$
$$= 0.8796.$$

(c) Given that $\cos \alpha = 0.8557$, find α.
We find in Table 2 that

$$\cos 31° = 0.8572 \quad \text{and} \quad \cos 31°30' = 0.8526.$$

Hence α lies between $31°$ and $31°30'$. We compute

$$\alpha = 31° + \frac{0.8572 - 0.8557}{0.8572 - 0.8526} \times 30'$$
$$= 31° + \frac{15}{46} \times 30'$$
$$= 31°10'.$$

The use of linear interpolation allows us to find the values of trigonometric functions that are usually accurate to four significant figures, except in case of the tangent function for angles greater than $75°$, for which linear interpolation does not give results of sufficient accuracy due to a rapid increase in the value of the function.
It is clear from Definition 1.1 that

(7) $\sin(90° - \alpha) = \cos \alpha,$

(8) $\cos(90° - \alpha) = \sin \alpha,$

(9) $\tan \alpha = \dfrac{\sin \alpha}{\cos \alpha}.$

Formulas (7) and (8) are immediate consequences of the fact that if $\angle BAC = \alpha$ and $\angle ACB = 90°$, then by (1), $\angle ABC = 90° - \alpha$. Formula (9) is obtained by dividing each side of (4) by the corresponding side of (5) and comparing the resulting equality with (6).
Before we show some of the applications of trigonometric functions, we shall use elementary geometry to compute the values of trigonometric functions for some special angles.
We first consider the angle of $30°$ (Figure 3.7). Let $\angle BAC = 30°$ and $\angle ABC = 90°$. Construct triangle ABD with $\angle BAD = 30°$ and $\angle ABD = 90°$. Then CBD is in a straight line. Moreover, the triangles ABC and ABD are congruent,

Figure 3.7

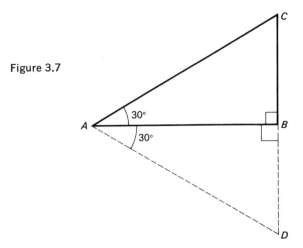

and therefore triangle CAD is isosceles with apex angle equal to $30° + 30° = 60°$. It follows that $\angle ACD = \angle ADC = 60°$ and that triangle CAD is in fact equilateral. Set $AC = b$. Then $AD = CD = b$ and $BC = BD = \frac{1}{2}b$. Also, by the Pythagorean theorem,

$$AB = \sqrt{AC^2 - BC^2}$$
$$= \sqrt{b^2 - \tfrac{1}{4}b^2}$$
$$= \frac{\sqrt{3}}{2}b.$$

Hence,

$$\sin 30° = BC/AC$$
$$= \frac{\frac{1}{2}b}{b}$$
$$= \tfrac{1}{2},$$

$$\cos 30° = AB/AC$$
$$= \frac{\frac{\sqrt{3}}{2}b}{b}$$
$$= \frac{\sqrt{3}}{2},$$

and

$$\tan 30° = \frac{\sin 30°}{\cos 30°}$$
$$= \frac{1/2}{\sqrt{3}/2}$$
$$= \frac{1}{\sqrt{3}}.$$

Now, using formulas (7), (8), and (9), we obtain

$$\sin 60° = \cos (90° - 60°)$$
$$= \tfrac{1}{2}\sqrt{3},$$
$$\cos 60° = \sin (90° - 60°)$$
$$= \tfrac{1}{2},$$
$$\tan 60° = \frac{\sin 60°}{\cos 60°}$$
$$= \sqrt{3}.$$

Now consider the angle of 45°. The triangle ABC in Figure 3.8 is right-angled with $\angle BAC = 45°$, and thus by elementary geometry the triangle is isosceles, i.e., $a = b$. It follows from the Pythagorean theorem that

$$c^2 = a^2 + b^2$$
$$= a^2 + a^2$$
$$= 2a^2,$$

i.e.,

$$c = \sqrt{2}\, a.$$

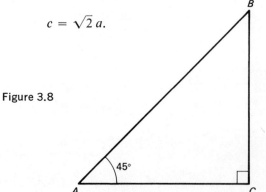

Figure 3.8

Hence,

$$\sin 45° = a/c$$
$$= \frac{a}{\sqrt{2}\, a}$$
$$= \frac{1}{\sqrt{2}},$$

$$\cos 45° = b/c$$
$$= a/c$$
$$= \frac{1}{\sqrt{2}},$$

and

$$\begin{aligned}
\tan 45° &= \frac{\sin 45°}{\cos 45°} \\
&= \frac{1/\sqrt{2}}{1/\sqrt{2}} \\
&= 1.
\end{aligned}$$

We can now put all of our computed values together in a single table.

(10)

FUNCTION	ANGLE		
	30°	45°	60°
sin	$\dfrac{1}{2}$	$\dfrac{1}{\sqrt{2}}$	$\dfrac{\sqrt{3}}{2}$
cos	$\dfrac{\sqrt{3}}{2}$	$\dfrac{1}{\sqrt{2}}$	$\dfrac{1}{2}$
tan	$\dfrac{1}{\sqrt{3}}$	1	$\sqrt{3}$

A useful mnemonic is that the sines of 30°, 45°, and 60° are $\frac{1}{2}\sqrt{1}$, $\frac{1}{2}\sqrt{2}$, and $\frac{1}{2}\sqrt{3}$, respectively.

We conclude this section with examples of applications of trigonometric functions. Further applications will be discussed in Chapter 5.

Example 1.3 If B is a point above the horizontal plane through a point A, then the *angle of elevation* of B at A is the angle between AB and the horizontal plane through A. If the point A is higher than B, then the appropriate term is *angle of depression*.

(a) The shadow of a vertical tree is 80 feet long when the angle of elevation of the sun is 20°. Find the height of the tree.

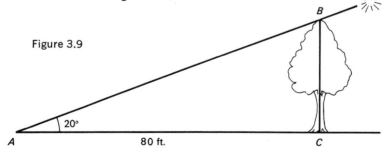

Figure 3.9

Let CB represent the tree and AC the shadow. Then $A = 20°$, $C = 90°$, $b = 80$ ft., and the problem is to find a.

We have

$$\tan 20° = a/b$$
$$= a/80,$$

and we find in Table 2 that $\tan 20° = 0.3640$. Hence,

$$a = 80 \tan 20°$$
$$= 80 \times 0.3640$$
$$= 29.12 \text{ ft.}$$

(b) An antenna pole is to be braced by three steel wires which are to be attached at a point on the pole 20 feet from the ground, and which are to be inclined 50° to the pole. Find the total length of steel wire necessary.

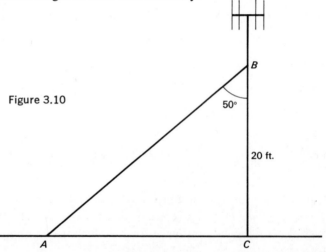

Figure 3.10

Let CB represent the lower twenty feet of the antenna pole and let AB represent one of the steel wires. Then in triangle ABC we have $C = 90°$, $B = 50°$, and $a = 20$ feet. Hence,

$$\cos B = \frac{a}{c},$$

i.e.,

$$c = \frac{a}{\cos B}$$
$$= \frac{20}{\cos 50°}.$$

We find in Table 2 that cos 50° = 0.6428, and we compute that c = 31.1 feet. Thus the total length of wire needed is 3 × 31.1 = 93.3 ft.

(c) A car traveling 60 m.p.h. due north crosses over a freeway at the same time as a truck which is traveling due west at 48 m.p.h. passes under the overpass. After t minutes the two vehicles are 10,000 feet apart (in a straight line). How far has the car travelled from the overpass?

Let C represent the overpass and let A and B represent the positions of the truck and the car, respectively, after t minutes (Figure 3.11).

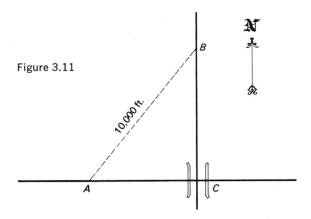

Figure 3.11

Then c = 10,000 feet, b = 48 kt, a = 60 kt where k = $\dfrac{5280}{60}$ is a numerical factor which converts m.p.h. to ft. per min. Hence

$$\begin{aligned} \tan A &= a/b \\ &= \frac{60\,kt}{48\,kt} \\ &= 1.25, \end{aligned}$$

and we find from Table 2 that A = 51°30′, approximately. Moreover,

$$\sin A = a/c,$$

and therefore

$$\begin{aligned} a &= c \sin A \\ &= 10{,}000 \sin 51°30′ \\ &= 10{,}000 \times 0.7826 \\ &= 7{,}826 \text{ feet.} \end{aligned}$$

SYNOPSIS

In this section we introduced the sine, cosine, and tangent functions of acute angles and discussed some simple properties of these functions. In particular, we computed the values of $\sin \alpha$, $\cos \alpha$, $\tan \alpha$ for $\alpha = 30°$, $45°$, and $60°$.

QUIZ

Answer *true* or *false:*
1. $0 < \sin \alpha < 1$ for any acute angle α.
2. $0 < \cos \alpha < 1$ for any acute angle α.
3. $0 < \tan \alpha < 1$ for any acute angle α.
4. $\tan (90° - \alpha) = \tan \alpha$ for any acute angle α.
5. $\sin 2\alpha = 2 \sin \alpha$ for any acute angle α.
6. If $\sin \alpha = 1/3$, then $\cos \alpha = 2/3$.
7. If $\sin \alpha = 1/3$, then $\cos (90° - \alpha) = 1/3$.
8. If $90° > \alpha > \beta > 0°$, then $\sin \alpha > \sin \beta$.
9. If $90° > \alpha > \beta > 0°$, then $\cos \alpha > \cos \beta$.
10. If $90° > \alpha > \beta > 0°$, then $\tan \alpha > \tan \beta$.

EXERCISES

1. Show that $\sin^2 \alpha + \cos^2 \alpha = 1$ for any acute angle α.
2. Prove that

$$1 + \tan^2 \alpha = \frac{1}{\cos^2 \alpha}$$

for any acute angle α.
3. Let ABC be a triangle right-angled at C. Find all angles and lengths of sides of ABC given that:

(a) $b = 3, c = 5$;   (b) $a = 12, c = 13$;

(c) $a = 1, b = 1$;   (d) $a = 1, b = \sqrt{3}$;

(e) $a = 1, c = 4$;   (f) $c = 3, A = 40°$;

(g) $a = 3, A = 40°$;   (h) $b = 3, A = 40°$;

(i) $a = 3.27, A = 49°30'$;   (j) $c = 15.6, A = 72°$;

(k) $b = 0.76, B = 20°$;   (l) $b = 2.4, A = 45°$;

(m) $c = 5, B = 48°50'$;   (n) $a = 100, A = 37°15'$.

4. The length of diagonal AC in rectangle $ABCD$ is 20 and $\angle BAC = 25°$. Find the lengths of the sides of the rectangle.
5. In a circle of radius 50 inches, a chord subtends an angle of $80°$ at the center. What is the length of the chord?

6. Point *A* is in the horizontal plane through *C*, the base of a tower. If the tower is 100 feet high and the angle of elevation of its top is 16° at *A*, how far is *A* from *C*?

7. Find the area of an isosceles triangle whose sides are 1 inch long and whose apex angle is 56°.

8. Find the area of a regular (a) pentagon, (b) hexagon, (c) octagon inscribed in a circle of radius 1 inch.

   [*Hint:* The area of an *n*-sided regular polygon inscribed in a circle of radius 1 inch is *n* times the area of the isosceles triangle whose sides are 1 inch long and whose apex angle is 360°/*n*.]

9. Find the area of a regular (a) pentagon, (b) hexagon, (c) octagon circumscribing a circle of radius 1 inch.

10. A ship sailing due south approaches pier *P*. Lighthouse *X* is situated 1,000 yards due west of *P*, and lighthouse *Y* is 1,000 yards due east of *P*. When the ship is at a point *A*, it is found that $\angle XAY = 50°$. One minute later the ship is at *B*, and it is found that $\angle XBY = 90°$. What is the speed of the ship in yards per minute?

    [*Hint:* Let the speed be *v* yards/min. Then $AB = v$ yards. Find *BP* from the right-angled triangle *XPB* and *AP* from the right-angled triangle *XPA*.]

11. What is the height of an antenna mast if, on walking 173 feet toward it on a horizontal line through its bottom, the angle of elevation of its top increases from 30° to 45°?

3.2 Functions of General Angles

In the preceding section we defined $\sin \alpha$, $\cos \alpha$, and $\tan \alpha$ for any acute angle α. Although these definitions suffice in some simple cases, many important applications of trigonometry require a generalized concept of angle and more general definitions of trigonometric functions.

Let l_1 and l_2 be two intersecting lines in the plane. As we learned in plane geometry, l_1 and l_2 define an *angle* θ: The *sides* of the angle are the two "half-lines" emanating from the point of intersection, i.e., the *vertex O*. We assign an orientation to the angle by thinking of the angle as a rotation from l_1 to l_2. The half-line l_1 is called the *initial side* of the angle θ, and the half-line l_2 is called the *terminal side* of θ.

Figure 3.12

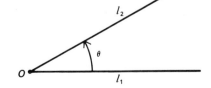

The curved arrow indicating θ designates the direction of rotation. Thus, there is another angle that can be associated with l_1 and l_2 as is illustrated in the following diagram.

Figure 3.13

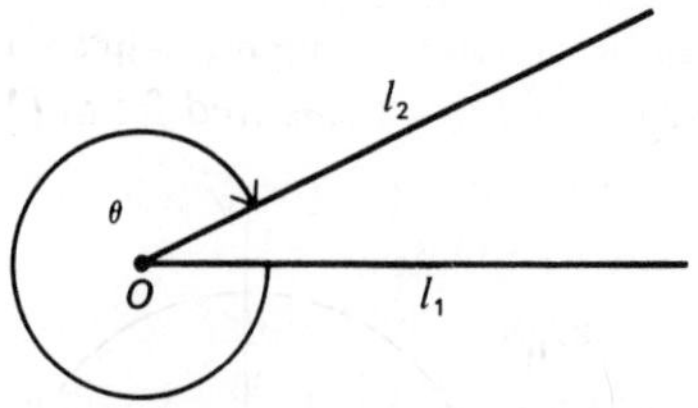

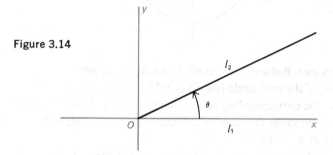

Any angle is congruent to an angle in *standard position* on the coordinate axes, i.e., an angle in which the initial side l_1 coincides with the nonnegative x-axis and the vertex O coincides with the origin.

Figure 3.14

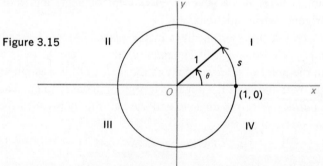

In trigonometry, we shall often assume that the angles which we discuss are in standard position.

There are two important measures for angles. Let θ be an angle in standard position. The terminal side of the angle cuts off an arc of length s on the *unit circle* (i.e., the circle of radius 1 and center at O) where the arc is measured counterclockwise from the point $(1, 0)$.

Figure 3.15

Clearly each angle θ uniquely defines an arc of length s, and each nonnegative number s corresponds to precisely one angle θ. We will systematically identify the angle θ with the length of the arc s and use the same notation (in this case θ)

for both. We indicate the direction of rotation of an angle θ by $+\theta$ when the rotation is in a counterclockwise direction (in this case the plus sign is usually omitted), and by $-\theta$ when the rotation is in a clockwise direction. In all cases we refer to arcs of the unit circle. Thus, $\theta = -3/4$ indicates a rotation in the clockwise direction through an arc of length 3/4 measured from (1, 0).

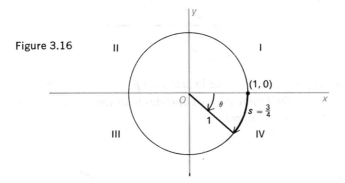

Figure 3.16

Definition 2.1 (Degree, Radian) An angle defined by an arc of length 1/360 of the circumference of the unit circle (which is 2π) is called a *degree*. In other words, 1 degree is the angle corresponding to an arc of length $2\pi/360$. This definition is equivalent to the definition of a degree in Section 3.1. An angle defined by an arc of length 1 is called a *radian*.

Since the circumference of the unit circle is 2π, it follows that an angle of 2π radians is the same as an angle of 360 degrees:

$$2\pi \text{ radians } = 360 \text{ degrees.}$$

Hence

(1) $$1 \text{ radian } = \frac{180}{\pi} \text{ degrees}$$

and

(2) $$1 \text{ degree } = \frac{\pi}{180} \text{ radians.}$$

If an angle is represented by a number without any additional notation, it is understood to be measured in radians. Recall that it is customary to denote degrees by inserting a zero superscript to the measure. Thus an angle θ of 25 degrees is written

$$\theta = 25°.$$

In other words, 25° corresponds to an arc of length

$$\frac{25 \cdot 2\pi}{360} = 25 \cdot \frac{\pi}{180} = \frac{25\pi}{180} = \frac{5\pi}{36}$$

measured counterclockwise on the unit circle, starting from the point $(1, 0)$.

An arc of length $2\pi/360$, i.e., a degree, is further divided into 60 arcs of equal length, each of them defining an angle which is called a *minute*. That is,

$$60 \text{ minutes} = 1°.$$

A minute is further divided into 60 equal parts called *seconds*. An angle of 1 second is very small indeed. It is roughly the angle subtended by a dime at a distance of $2\frac{1}{4}$ miles. A minute is usually denoted by a superscript dash, and a second with a superscript double dash. Thus an angle θ of 36 degrees 15 minutes 18 seconds is written

$$\theta = 36°15'18''.$$

In this case, θ corresponds to an arc measured counterclockwise on the unit circle from the point $(1, 0)$ and of length

$$36 \cdot \frac{2\pi}{360} + 15 \cdot \frac{2\pi}{60 \cdot 360} + 18 \cdot \frac{2\pi}{60 \cdot 60 \cdot 360}.$$

Example 2.1   (a)   What is the radian measure of an angle of 90°?
Since 1 degree is $\pi/180$ radians, it follows that

$$90° = 90 \cdot \pi/180 \text{ radians}$$
$$= \pi/2 \text{ radians.}$$

(b)   Express an angle of 4π radians in terms of degrees.
In this case, since 2π radians $= 360°$, it follows that

$$4\pi \text{ radians} = 720°.$$

(c)   Express in degrees: $\pi, \pi/2, \pi/3, \pi/4, \pi/6, 0, 1, -\pi/2$ radians.
By (1), these angles are equal to $180°, 90°, 60°, 45°, 30°, 0°, 180°/\pi = (57.29\ldots)°$, $-90°$, respectively.

(d)   Express in radians: $1°, 120°, -22.5°, 720°, -810°$.
By (2), these angles are equal to $\pi/180 = 0.0174\ldots, 2\pi/3, -\pi/8, 4\pi, -9\pi/2$ radians, respectively.

We are now ready to define the first two standard trigonometric functions.

Definition 2.2 (Sine and Cosine) Let α be an angle (either in radians or in degrees) in standard position. Let the terminal side of α cut the unit circle at the point (x, y). Then the *sine function* and the *cosine function*, abbreviated *sin* and *cos*, respectively, are real-valued functions defined on the set of angles by

$$\sin \alpha = y,$$
$$\cos \alpha = x.$$

For example,

$$\sin 0° = 0, \qquad \cos 0° = 1;$$
$$\sin 90° = 1, \qquad \cos 90° = 0;$$
$$\sin 180° = 0, \qquad \cos 180° = -1;$$
$$\sin 270° = -1, \quad \cos 270° = 0;$$
$$\sin 360° = 0, \qquad \cos 360° = 1.$$

Observe that the values of the sine or the cosine function depend only on the terminal side of the angle. In other words, the values of sin α and cos α depend only on the final position of the rotation, not on its previous history. Thus two angles that differ by complete revolutions of 360° have the same sine values and the same cosine values. Hence we have

(3) $$\sin (\alpha + k \cdot 360°) = \sin \alpha$$

and

(4) $$\cos (\alpha + k \cdot 360°) = \cos \alpha,$$

for any angle of α degrees and any integer k. We say that the sine and the cosine functions are *periodic*, with *period* 360° (or 2π radians). This simply means that each of these functions repeats itself every 360°.

The domain of both functions in Definition 2.2 is the set of all angles. We use the same notation as for the trigonometric functions defined in Section 3.1, because for angles between 0° and 90° the definitions are equivalent. In fact, if α is acute, let P be the point (x, y) and let A be the foot of the perpendicular from P to the x-axis.

Figure 3.17

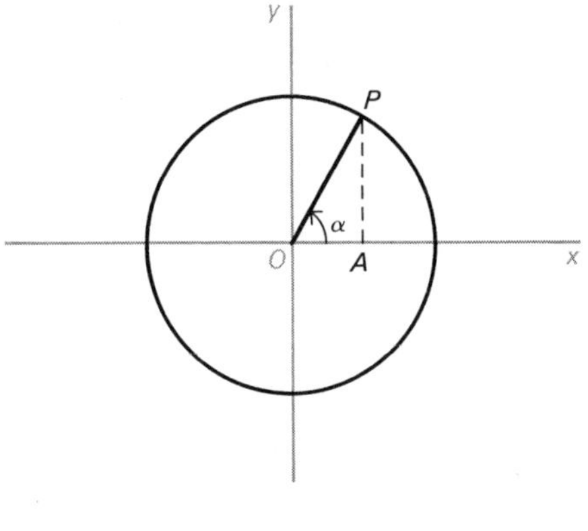

Then from triangle OAP, right-angled at A, we have according to Definition 1.1,

$$\sin \alpha = \frac{AP}{OP}$$
$$= \frac{y}{1}$$
$$= y.$$

Similarly,

$$\cos \alpha = \frac{OA}{OP}$$
$$= \frac{x}{1}$$
$$= x.$$

These are precisely the values at the angle α of the sine and the cosine functions in Definition 2.2.

Example 2.2   Find all angles α for which $\sin \alpha = 0$ and all angles β for which $\cos \beta = 0$.

By the definition of the sine function, $\sin \alpha = 0$ if and only if the terminal side of the angle lies along the x-axis. Thus, for nonnegative angles less than $360°$, $\sin \alpha = 0$ implies $\alpha = 0°$ or $180°$. It follows from (3) that, in general, $\sin \alpha = 0$ if and only if α is an integral multiple of $180°$. Similarly, we find that $\cos \beta = 0$ if and only if

$$\beta = 90° + k \cdot 180°$$

for some integer k.

We shall now define the other four standard trigonometric functions.

Definition 2.3 (Tangent, Cotangent, Secant, Cosecant)   If $\cos \alpha \neq 0$, then the *tangent* of α, written tan α, is defined by

(5)
$$\tan \alpha = \frac{\sin \alpha}{\cos \alpha},$$

and the *secant* of α, written sec α, is defined by

(6)
$$\sec \alpha = \frac{1}{\cos \alpha}.$$

If $\sin \alpha \neq 0$, then the *cotangent* of α, written $\cot \alpha$, is defined by

(7)
$$\cot \alpha = \frac{\cos \alpha}{\sin \alpha},$$

and the *cosecant* of α, written $\operatorname{cosec} \alpha$, is defined by

$$\operatorname{cosec} \alpha = \frac{1}{\sin \alpha}.$$

Thus tangent and secant are real-valued functions whose domain is the set of all angles α for which $\cos \alpha \neq 0$. Cotangent and cosecant are real-valued functions whose domain is the set of all angles α for which $\sin \alpha \neq 0$, i.e., the set of all angles α that are not integer multiples of $180°$ (see Example 2.2).

Let α be an angle in standard position. Let the terminal side of α cut the unit circle at a point $P = (x, y)$ and let $R = (1, 0)$ and $O = (0, 0)$.

Figure 3.18

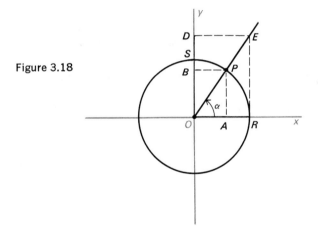

We shall give a geometric interpretation of our definition of the trigonometric functions. Draw a line through P parallel to the y-axis and intersecting the x-axis at A. Draw another line through P parallel to the x-axis and intersecting the y-axis at B. Then, by Definition 2.2,

$$\sin \alpha = OB$$
and
$$\cos \alpha = OA,$$

where, as usual, OB denotes the length of the segment OB if B is above the x-axis, and minus that length if B is below the x-axis. Similarly, OA denotes the length of the segment OA or minus that length according as A is respectively to the right or to the left of the y-axis. Now, draw the line which is tangent to the unit circle at R; let it intersect the straight line containing OP at point E. Then, by elementary geometry, the triangles ORE and OAP have corresponding angles equal and thus are similar. Hence,

$$
\begin{aligned}
\tan \alpha &= \frac{\sin \alpha}{\cos \alpha} \\
&= AP/OA \\
&= RE/OR \\
&= RE,
\end{aligned}
$$

where again, RE is the length of the segment RE or minus that length, according as E is above or below the x-axis. In fact, E is above the x-axis when α is in the first or third quadrant, in which case $\tan \alpha = \sin \alpha/\cos \alpha$ is positive; and E is below the x-axis when α is in the second or fourth quadrant, in which case $\tan \alpha = \sin \alpha/\cos \alpha$ is negative.

It is clear from the definitions of the trigonometric functions and from the above geometric representation that the sign of a trigonometric function depends only on the quadrant in which the terminal side of the angle is situated. The following table gives the sign of each function in each quadrant.

FUNCTION	QUADRANT			
	I	II	III	IV
Sine	+	+	−	−
Cosine	+	−	−	+
Tangent	+	−	+	−
Cotangent	+	−	+	−
Secant	+	−	−	+
Cosecant	+	+	−	−

Actually, even more specific information can be obtained from the geometric representation of trigonometric functions. In the following theorem, the angle α is given in degrees. Clearly all parts of the theorem can be expressed equivalently in radians. Also, the theorem is stated in terms of the sine and the cosine functions. The corresponding theorems for the other trigonometric functions follow immediately from Definition 2.3 (see Example 2.3).

Theorem 2.1 *Let α be any angle. Then*

$$\text{(8)} \qquad\qquad \sin(-\alpha) = -\sin\alpha,$$

$$\text{(9)} \qquad\qquad \cos(-\alpha) = \cos\alpha,$$

$$\text{(10)} \qquad\qquad \sin(180° - \alpha) = \sin\alpha,$$

$$\text{(11)} \qquad\qquad \cos(180° - \alpha) = -\cos\alpha,$$

$$\text{(12)} \qquad\qquad \sin(90° - \alpha) = \cos\alpha,$$

$$\text{(13)} \qquad\qquad \cos(90° - \alpha) = \sin\alpha.$$

Proof The angle $-\alpha$ represents a rotation of the same amount as angle α, except that the rotation of $-\alpha$ degrees is in the direction opposite to that of α degrees. That is, if α is a clockwise rotation then $-\alpha$ is counterclockwise, and vice versa. Since the initial sides are the same when α and $-\alpha$ are in standard position, the terminal sides must be symmetrically situated with respect to the x-axis. Hence, if (x, y) is the point where the terminal side of angle α intersects the unit circle, then $(x, -y)$ is the point where the terminal side of $-\alpha$ cuts the circle. Thus, by Definition 2.2,

$$\sin(-\alpha) = -y$$
$$= -\sin\alpha,$$

while

$$\cos(-\alpha) = x$$
$$= \cos\alpha.$$

In order to prove (10) and (11), suppose first that $0° < \alpha < 90°$. Then α and $180° - \alpha$ are supplementary (i.e., their sum is $180°$), and thus their terminal sides are symmetric with respect to the y-axis. Therefore if α is in standard position and if $P = (x, y)$ is the point of intersection of the terminal side of α and the unit circle, then $Q = (-x, y)$ is the corresponding intersection point for the angle $180° - \alpha$. This is particularly easy to see from Figure 3.19 on the next page. By elementary geometry, the triangles OMP and ONQ are congruent. Thus, (10) and (11) follow in this case. If $90° < \alpha < 180°$, a similar argument, with the roles of P and Q interchanged, yields the same equalities. Clearly the equalities hold for the special values $\alpha = 0°, 90°$, and $180°$. If $-180° < \alpha < 0°$, then by (3) and (8),

$$\begin{aligned}
\text{(14)} \qquad \sin(180° - \alpha) &= \sin\left(-(180° + \alpha) + 360°\right) \\
&= \sin\left(-(180° + \alpha)\right) \\
&= -\sin(180° + \alpha) \\
&= -\sin\left(180° - (-\alpha)\right).
\end{aligned}$$

Figure 3.19

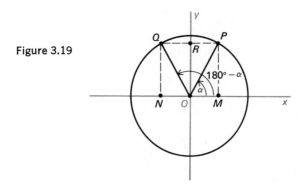

Now, $0 < -\alpha < 180°$. Hence

(15) $$\sin\left(180° - (-\alpha)\right) = \sin\left(-\alpha\right),$$

and by (8),

(16) $$\sin\left(-\alpha\right) = -\sin\alpha.$$

Putting (14), (15), and (16) together we get (10). Hence, we have proved (10) for all α between $-180°$ and $180°$. Now, we are ready to prove (10) for any angle α. Dividing α by 360 we can find an integer k such that

$$\alpha = \beta + k \cdot 360°,$$

where β is an angle satisfying $-180° \le \beta \le 180°$. Therefore, by (3),

$$
\begin{aligned}
\sin\alpha &= \sin\left(\beta + k \cdot 360°\right) \\
&= \sin\beta \\
&= \sin\left(180° - \beta\right) \\
&= \sin\left((180° - \beta) - k \cdot 360°\right) \\
&= \sin\left(180° - (\beta + k \cdot 360°)\right) \\
&= \sin\left(180° - \alpha\right).
\end{aligned}
$$

Formula (11) is proved similarly.

To prove formulas (12) and (13), suppose first that $0° < \alpha < 90°$. Let α and $90° - \alpha$ be in standard position, and let their terminal sides intersect the unit circle at points P and Q, respectively. (See Figure 3.20.) Let QM and PN be the two lines parallel to the y-axis which intersect the x-axis at M and N, respectively. Draw QT and PS parallel to the x-axis intersecting the y-axis at T and S, respectively.

Figure 3.20

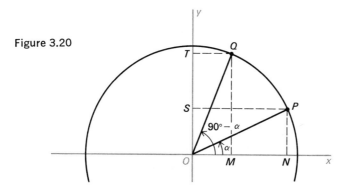

Now, the right-angled triangles ONP and OTQ are congruent since $OP = OQ = 1$ and $\angle NOP = \angle QOT = \alpha$. Thus, $NP = TQ$ and $ON = OT$. Hence,

$$\begin{aligned}
\sin (90° - \alpha) &= OT \\
&= ON \\
&= \cos \alpha.
\end{aligned}$$

Similarly,

$$\begin{aligned}
\cos (90° - \alpha) &= OM \\
&= TQ \\
&= NP \\
&= OS \\
&= \sin \alpha.
\end{aligned}$$

Next, assume that $90° < \alpha < 180°$. Then $0° < \alpha - 90° < 90°$ and

$$\begin{aligned}
\sin (90° - \alpha) &= -\sin (\alpha - 90°) \\
&= -\cos \left(90° - (\alpha - 90°)\right) \\
&= -\cos (180° - \alpha) \\
&= \cos \alpha,
\end{aligned}$$

by (11). Similarly,

$$\begin{aligned}
\cos (90° - \alpha) &= \cos (\alpha - 90°) \\
&= \sin \left(90° - (\alpha - 90°)\right) \\
&= \sin (180° - \alpha) \\
&= \sin \alpha.
\end{aligned}$$

It is trivial to verify that formulas (12) and (13) hold for $\alpha = 0°$, $90°$, and $180°$. Further, if $-180° < \alpha < 0°$, then $0° < -\alpha < 180°$, and

$$\begin{aligned}
\sin (90° - \alpha) &= \sin \big(180° - (90° + \alpha)\big) \\
&= \sin \big(90° - (-\alpha)\big) \\
&= \cos (-\alpha) \\
&= \cos \alpha.
\end{aligned}$$

Also in this case,

$$\begin{aligned}
\cos (90° - \alpha) &= \cos \big(180° - (90° + \alpha)\big) \\
&= -\cos \big(90° - (-\alpha)\big) \\
&= -\sin (-\alpha) \\
&= \sin \alpha.
\end{aligned}$$

Lastly, let α be any angle. Then there exist an integer k and an angle β, $-180° \le \beta \le 180°$, such that

$$\alpha = \beta + k \cdot 360°.$$

We have shown above that formulas (12) and (13) hold for angles between $-180°$ and $180°$. Thus,

$$\begin{aligned}
\sin (90° - \alpha) &= \sin (90° - \beta - k \cdot 360°) \\
&= \sin (90° - \beta) \\
&= \cos \beta \\
&= \cos (\beta + k \cdot 360°) \\
&= \cos \alpha.
\end{aligned}$$

Formula (13) can be proved in a similar way, or from (12),

$$\begin{aligned}
\cos (90° - \alpha) &= \sin \big(90° - (90° - \alpha)\big) \\
&= \sin \alpha.
\end{aligned}$$

This concludes the proof of Theorem 2.1. ∎

Example 2.3   Extend Theorem 2.1 to the other trigonometric functions.

If α lies in the domain of the function on the left-hand side of each of the formulas, we have

(17)
$$\begin{aligned}
\tan (-\alpha) &= \frac{\sin (-\alpha)}{\cos (-\alpha)} \\
&= \frac{-\sin \alpha}{\cos \alpha} \\
&= -\tan \alpha;
\end{aligned}$$

$$\cot(-\alpha) = \frac{\cos(-\alpha)}{\sin(-\alpha)}$$

(18)
$$= \frac{\cos\alpha}{-\sin\alpha}$$

$$= -\cot\alpha;$$

$$\sec(-\alpha) = \frac{1}{\cos(-\alpha)}$$

(19)
$$= \frac{1}{\cos\alpha}$$

$$= \sec\alpha;$$

$$\csc(-\alpha) = \frac{1}{\sin(-\alpha)}$$

(20)
$$= \frac{1}{-\sin\alpha}$$

$$= -\csc\alpha;$$

$$\tan(180° - \alpha) = \frac{\sin(180° - \alpha)}{\cos(180° - \alpha)}$$

(21)
$$= \frac{\sin\alpha}{-\cos\alpha}$$

$$= -\tan\alpha;$$

$$\cot(180° - \alpha) = \frac{\cos(180° - \alpha)}{\sin(180° - \alpha)}$$

(22)
$$= \frac{-\cos\alpha}{\sin\alpha}$$

$$= -\cot\alpha;$$

$$\sec(180° - \alpha) = \frac{1}{\cos(180° - \alpha)}$$

(23)
$$= \frac{1}{-\cos\alpha}$$

$$= -\sec\alpha;$$

$$\csc(180° - \alpha) = \frac{1}{\sin(180° - \alpha)}$$

(24)
$$= \frac{1}{\sin\alpha}$$

$$= \csc\alpha;$$

$$\tan (90° - \alpha) = \frac{\sin (90° - \alpha)}{\cos (90° - \alpha)}$$

(25)
$$= \frac{\cos \alpha}{\sin \alpha}$$

$$= \cot \alpha;$$

(26)
$$\cot (90° - \alpha) = \tan \left(90° - (90° - \alpha)\right)$$
$$= \tan \alpha;$$

$$\sec (90° - \alpha) = \frac{1}{\cos (90° - \alpha)}$$

(27)
$$= \frac{1}{\sin \alpha}$$

$$= \operatorname{cosec} \alpha;$$

(28)
$$\operatorname{cosec} (90° - \alpha) = \sec \left(90° - (90° - \alpha)\right)$$
$$= \sec \alpha.$$

Some important identities are obtained directly from the Pythagorean theorem or, equivalently, from the distance formula. Denote $(\sin \alpha)^2$ and $(\cos \alpha)^2$ by $\sin^2 \alpha$ and $\cos^2 \alpha$, respectively, and similarly for the other trigonometric functions.

Theorem 2.2   *If α is any angle, then*

(29)
$$\sin^2 \alpha + \cos^2 \alpha = 1.$$

If $\cos \alpha \neq 0$, then

(30)
$$\tan^2 \alpha + 1 = \sec^2 \alpha.$$

If $\sin \alpha \neq 0$, then

(31)
$$\cot^2 \alpha + 1 = \operatorname{cosec}^2 \alpha.$$

Proof   Let α be in standard position, and let P be the point of intersection of the terminal side of α and the unit circle. Then, by Definition 2.2,

$$P = (\cos \alpha, \sin \alpha)$$

and the square of the length of OP (i.e., the distance from O to P) is

$$OP^2 = \cos^2 \alpha + \sin^2 \alpha.$$

But *OP* is the radius of the unit circle, and thus

$$1 = \cos^2 \alpha + \sin^2 \alpha,$$

which is the identity (29).

Now, if $\cos \alpha \neq 0$, divide both sides of (29) by $\cos^2 \alpha$ to obtain

$$\frac{\sin^2 \alpha}{\cos^2 \alpha} + 1 = \frac{1}{\cos^2 \alpha},$$

which is (30) by Definition 2.3. Identity (31) is obtained similarly. ∎

Example 2.4 If $\sin \alpha = 3/5$ and $0° < \alpha < 90°$, find $\cos \alpha$ and $\tan \alpha$.

By formula (29),

$$\begin{aligned}
\cos^2 \alpha &= 1 - \sin^2 \alpha \\
&= 1 - (3/5)^2 \\
&= 16/25.
\end{aligned}$$

Since α is in the first quadrant, $\cos \alpha > 0$. Thus,

$$\begin{aligned}
\cos \alpha &= \sqrt{16/25} \\
&= 4/5.
\end{aligned}$$

It follows immediately that

$$\tan \alpha = 3/4.$$

SYNOPSIS

The general notion of an angle in standard position was defined, and the degree and radian measures of an angle were explained. The trigonometric functions $\sin \alpha$, $\cos \alpha$, $\tan \alpha$, $\cot \alpha$, $\sec \alpha$, $\csc \alpha$ were introduced in Definitions 2.2 and 2.3. Various relations between these functions were derived in Theorem 2.1. Finally, in Theorem 2.2, the Pythagorean theorem was restated in terms of the trigonometric functions.

QUIZ

Answer *true* or *false:*

1. If $\alpha = 1.5$ radians, then α is an obtuse angle (i.e., an angle in standard position whose terminal side is in the second quadrant).

2. $\dfrac{\csc \alpha}{\sec \alpha} = \tan \alpha$ for any angle α for which all three functions are defined.

3. If $\alpha = \pi/4$ radians, then $\tan \alpha = \cot \alpha$.
4. $\sin (90° + \alpha) = \cos \alpha$ for all α.
5. $\cos (90° + \alpha) = \sin \alpha$ for all α.
6. $\sin (180° + \alpha) = -\sin \alpha$ for all α.
7. $\cos (180° + \alpha) = -\cos \alpha$ for all α.
8. $\tan (k \cdot 180°) = 0$ for all integers k.
9. $\operatorname{cosec}^2 \alpha + \sec^2 \alpha = 1$ for all α for which both functions are defined.
10. If $\beta = \pi/3$ radians, then $\sin (2\beta) = \cos (\tfrac{1}{2}\beta)$.

EXERCISES

1. Find the radian measure of each of the following angles given in degrees:
   (a) 0°,   (b) 720°,   (c) 180°,
   (d) −90°,   (e) −10°,   (f) −360°,
   (g) 45°,   (h) 270°,   (i) 30°,
   (j) −45°.

2. Express the following angles in terms of degrees:
   (a) 0 radians,   (b) -2π radians,   (c) $3\pi/2$ radians,
   (d) $2\pi/3$ radians,   (e) π radians,   (f) $\pi/12$ radians,
   (g) $3\pi/5$ radians,   (h) 3 radians,   (i) $-\tfrac{1}{2}$ radians,
   (j) π^2 radians.

3. Find the values of the sine, cosine, and tangent functions (whenever defined) for each of the following angles:
   (a) $\pi/2$,   (b) $-3\pi/2$,   (c) 4π,
   (d) $\pi/6$,   (e) $-\pi/6$,   (f) $\pi/3$,
   (g) $-\pi/3$,   (h) π,   (i) $-\pi$,
   (j) $13\pi/6$.

4. Given that $\sin 32° = 0.53$ (correct to two decimal places), evaluate:
   (a) $\cos 32°$,   (b) $\tan 32°$,   (c) $\cot 32°$,
   (d) $\sec 32°$,   (e) $\operatorname{cosec} 32°$,   (f) $\sin 58°$,
   (g) $\sec 58°$,   (h) $\sin 148°$,   (i) $\tan 148°$,
   (j) $\cos (-32°)$,   (k) $\sin 122°$,   (l) $\sin 212°$,
   (m) $\sin (-752°)$,   (n) $\sin (-688°)$.

5. Let α be any angle for which $\tan \alpha$ is defined. Show that

$$\tan (\alpha + k \cdot 180°) = \tan \alpha$$

   for any integer k.

6. Find all angles x which satisfy the following equations:
   (a) $\sin x = \tfrac{1}{2}$;   (b) $\cos x = \dfrac{\sqrt{3}}{2}$;

(c) $\tan x = 1$;   (d) $\sin 2x = 0$;
(e) $\cos 2x = 0$;   (f) $\cos 3x = 1$;
(g) $\sin (45° + x) = \frac{1}{2}$;   (h) $\tan (30° - x) = \sqrt{3}$;
(i) $\sqrt{3} \sin x = \cos x$;   (j) $\tan x = \cot x$.

7. Let α be an angle in the first quadrant, $0° < \alpha < 90°$. Show that

$$\tan \alpha > \sin \alpha.$$

8. Show that in any triangle whose angles are α, β, γ,

$$\sin (\alpha + \beta) = \sin \gamma,$$
$$\cos (\alpha + \beta) = -\cos \gamma.$$

[*Hint:* The sum of the angles in a triangle is 180°.]

3.3 Trigonometric Tables

In Section 3.1 we explained how to use Table 2 of the Appendix to find the values of the sine, cosine and tangent of any acute angle. The cosecant, secant, and cotangent of an angle are obtained by computing the reciprocals of the sine, cosine and tangent of the angle, respectively. They can also be computed by the use of appropriate trigonometric formulas. For example,

$$\cot 24°30' = \tan (90° - 24°30')$$
$$= \tan 65°30'$$
$$= 2.194.$$

If the angle is not between 0° and 90°, then by adding a suitable multiple of 360° we can find an angle between $-180°$ and 180° whose sine, cosine, or tangent is equal to that of the original angle. Using formulas (8), (9), (10), (11), (17), and (21) in Section 3.2, we can then reduce the problem to one of finding the sine, cosine, or tangent of an angle between 0° and 90°.

Example 3.1 (a) Using the tables, evaluate $\sin 700°$, $\cos 700°$, $\tan 700°$.

$$\sin 700° = \sin (2 \times 360° - 20°)$$
$$= \sin (-20°)$$
$$= -\sin 20°$$
$$= -0.3420;$$

$$\cos 700° = \cos (-20°)$$
$$= \cos 20°$$
$$= 0.9397;$$
$$\tan 700° = \tan (-20°)$$
$$= -\tan 20°$$
$$= -0.3640.$$

(b) Using the tables, evaluate sin $(-950°)$, cos $(-950°)$, and tan $(-950°)$.

$$\sin (-950°) = \sin (3 \times 360° - 950°)$$
$$= \sin 130°$$
$$= \sin (180° - 50°)$$
$$= \sin 50°$$
$$= 0.7660;$$
$$\cos (-950°) = \cos 130°$$
$$= \cos (180° - 50°)$$
$$= -\cos 50°$$
$$= -0.6428;$$
$$\tan (-950°) = \tan 130°$$
$$= \tan (180° - 50°)$$
$$= -\tan 50°$$
$$= -1.192;$$

or alternatively, using the result in Exercise 5 of the preceding section

$$\tan (-950°) = \tan (5 \times 180° - 950°)$$
$$= \tan (-50°)$$
$$= -\tan 50°$$
$$= -1.192.$$

Trigonometric functions occur in many formulas in applied mathematics. Evaluation of these may often be facilitated by the use of logarithms. The method is essentially the same as the one used in Section 2.5. In order to shorten the task, however, it is convenient to have tables of logarithms of trigonometric functions (Table 3 of the Appendix). Note that for typographical reasons the bar notation is not used for negative characteristic in these tables. Otherwise, the tables are quite simple to use as will be seen from the following examples.

Example 3.2 Use logarithms to evaluate the following:

(*i*) $\qquad\qquad a = 643 \sin 37°;$

(ii) $$b = \frac{26.7}{\cos 36°30'} ;$$

(iii) $c = 0.243 \sin 31° \cos 85°;$

(iv) $$d = \frac{341 \tan 47°}{\cos 23°30'} ;$$

(v) $e = 25.3 \cos 41° \times 6.09 \cos 32°;$

(vi) $f = 25.3 \cos 41° + 6.09 \cos 32°;$

(vii) $$g = \frac{52 \tan 26° \times 372 \cos 48°}{24.3 \sin 74°} .$$

(i)

Numbers	Logs	Antilogs
643	2.8082	
× sin 37°	+ $\bar{1}$.7795	
	2.5877	387 = a

(ii)

Numbers	Logs	Antilogs
26.7	1.4265	
÷ cos 36°30′	− $\bar{1}$.9052	
	1.5213	33.2 = b

(iii)

Numbers	Logs	Antilogs
0.243	$\bar{1}$.3856	
× sin 31°	+ $\bar{1}$.7118	
× cos 85°	+ $\bar{2}$.9403	
	$\bar{2}$.0377	0.0109 = c

(iv)

Numbers	Logs	Antilogs
341	2.5328	
× tan 47°	+ 0.0303	
	2.5631	
÷ cos 23°30′	− $\bar{1}$.9624	
	2.6007	399 = d

(v)

Numbers	Logs	Antilogs
25.3	1.4031	
× cos 41°	+ $\bar{1}$.8778	
× 6.09	+ 0.7846	
× cos 32°	+ $\bar{1}$.9284	
	1.9939	98.6 = e

(*vi*)

Numbers	Logs	Antilogs
25.3	1.4031	
× cos 41°	+ $\bar{1}$.8778	
	1.2809	19.1
6.09	0.7846	
× cos 32°	+ $\bar{1}$.9284	
	0.7130	+ 5.16
		24.26 = *f*

[Note carefully the difference between computations (*v*) and (*vi*).]

(*vii*)

Numbers	Logs		Antilogs
52	1.7160		
× tan 26°	+ $\bar{1}$.6882		
× 372	+ 2.5705		
× cos 48°	+ $\bar{1}$.8255		
		3.8002	
÷ { 24.3	1.3856		
{ × sin 74°	+ $\bar{1}$.9828		
		− 1.3684	
		2.4318	270 = *g*

If either the angle or the logarithm of a value of a trigonometric function or of an antilogarithm is not in the tables, we can resort to linear interpolation. The method is analogous to the one used in Sections 2.5 and 3.1, and is illustrated in the following example.

Example 3.3 Evaluate the following to four significant figures. Use linear interpolation if necessary.

(a) $p = 34.28 \sin 126°25'$;

(b) $q = \dfrac{246.3 \cos 110°}{\tan 155°15'}$;

(c) $r = 616.3 \sin 35°20' \times 24.57 \tan 80°30'$;

(d) $s = 616.3 \sin 35°20' + 24.57 \tan 80°30'$;

(e) $t = 2.635 \sec 322°20'$;

(f) $u = \cot 26°50' \times \operatorname{cosec} 136°40'$.

(a) We first evaluate:

$$\log 34.28 = \log 34.20 + 0.8 \,(\log 34.30 - \log 34.20)$$
$$= 1.5340 + 0.8 \times 0.0013$$
$$= 1.5350;$$

$$\log \sin 126°25' = \log \sin 53°35'$$
$$= \log \sin 53°30' + \frac{5}{30} \,(\log \sin 54° - \log \sin 53°30')$$
$$= \overline{1}.9052 + \frac{5}{30} \times 0.0028$$
$$= \overline{1}.9057.$$

Numbers	Logs	Antilogs
34.28	1.5350	
$\times \sin 126°25'$	$+\ \overline{1}.9057$	
	1.4407	$27.59 = p$

For,

$$\text{antilog } 1.4407 = \text{antilog } 1.4393 + \frac{14}{16} \times 0.1$$
$$= 27.59.$$

[In the remaining examples we omit the details of linear interpolation since these computations can easily be performed mentally.]

(b)

Numbers		Logs	Antilogs
246.3	246.3	2.3914	
$\times \cos 110°$	$\times\,(-1)\cos 70°$	$+\ \overline{1}.5341$	
		1.9255	
$\div \tan 155°15'$	$\div\,(-1)\tan 24°45'$	$-\ \overline{1}.6637$	
		2.2618	$182.7 = q$

Note that

$$\cos 110° = \cos (180° - 70°) = -\cos 70°$$
$$\tan 155°15' = \tan (180° - 24°45') = -\tan 24°45',$$

but, of course, $\overline{1}.5341 = \log \cos 70°$, not the logarithm of $-\cos 70°$ which is not in the domain of the logarithm function. Neither is $\overline{1}.6637$ (nor even $-\overline{1}.6637$) the logarithm of $-\tan 24°45'$.

(c)

Numbers	Logs	Antilogs
616.3	2.7898	
× sin 35°20′	+ $\overline{1}$.7622	
× 24.57	+ 1.3904	
× tan 80°30′	+ 0.7764	
	4.7188	52340 = r

(d)

Numbers	Logs	Antilogs
616.3	2.7898	
× sin 35°20′	+ $\overline{1}$.7622	
	2.5520	356.5
× 24.57	1.3904	
× tan 80°30′	+ 0.7764	
	2.1668	+ 146.8
		503.3 = s

(e)

$$\sec 322°20' = 1/\cos 322°20'$$
$$= 1/\cos 37°40'.$$

Numbers	Logs	Antilogs
2.635	0.4208	
÷ cos 37°40′	− $\overline{1}$.8985	
	0.5223	3.329 = t

Note that log cos 37°30′ = $\overline{1}$.8995 and log cos 38° = $\overline{1}$.8965. Therefore we interpolate as follows:

$$\log \cos 37°40' = \overline{1}.8995 + \frac{10}{30}(-0.0030)$$
$$= \overline{1}.8985.$$

(f)

$$\cot 26°50' = \tan 63°10',$$
$$\operatorname{cosec} 136°40' = 1/\sin 136°40'$$
$$= 1/\sin 43°20'.$$

Numbers	Logs	Antilogs
tan 63°10′	0.2960	
÷ sin 43°20′	− $\overline{1}$.8365	
	0.4595	2.881 = u

SYNOPSIS

In this section we showed how to use trigonometric and logarithm tables in computations involving trigonometric functions.

QUIZ

Answer *true* or *false:*
1. $\cos \alpha = \sin (270° + \alpha)$, for all α.
2. $\sin \alpha = 1/\cos \alpha$, for all α for which both sides of the equality are defined.
3. $\tan \alpha = 1/\cot \alpha$, for all α for which both sides of the equality are defined.
4. $\tan^2 \alpha = 1 - \cot^2 \alpha$, for all α for which both sides of the equality are defined.
5. $\sin (45° + \alpha) = \cos (45° - \alpha)$, for all angles α.
6. $\sin \alpha + \sin \beta = \sin (\alpha + \beta)$, for all α and β.
7. $\log (\sin \alpha + \sin \beta) = \log \sin \alpha + \log \sin \beta$, for all α and β for which both sides of the equality are defined.
8. $\log (\sin \alpha \cdot \sin \beta) = \log \sin \alpha + \log \sin \beta$, for all α and β for which both sides of the equality are defined.
9. The characteristic of $\log \sin \alpha$ is $\bar{1}$, for all acute angles α (i.e., $0° < \alpha < 90°$).
10. If α is an acute angle, then $\sin \alpha < \tan \alpha$.

EXERCISES

1. Use suitable formulas and Table 2 to find the value of:
   (a) $\sin 325°$,   (b) $\cos (-325°)$,   (c) $\tan 110°$,
   (d) $\cot 110°$,   (e) $\sin 200°$,   (f) $\cos 200°$,
   (g) $\tan 200°$,   (h) $\cot 200°$,   (i) $\sec 200°$,
   (j) $\operatorname{cosec} 200°$,   (k) $\sin 117°30'$,   (l) $\cos 236°30'$,
   (m) $\tan 90°30'$,   (n) $\tan (-90°30')$,   (o) $\sin 997°$,
   (p) $\cos (-997°)$,   (q) $\sin 181°30'$,   (r) $\cos 181°30'$,
   (s) $\tan 181°30'$.

2. Use logarithms (Tables 1 and 3) to compute the value of $S = \frac{1}{2}xy \sin \theta$, given that:
   (a) $x = 3.15, y = 4.73, \theta = 46°$;
   (b) $x = 0.46, y = 0.238, \theta = 73°$;
   (c) $x = 248, y = 96, \theta = 100°$;
   (d) $x = y = 85.6, \theta = 37°30'$;
   (e) $x = 0.0631, y = 0.00937, \theta = 103°30'$;
   (f) $x = 6.53, y = 0.893, \theta = 143°30'$.

3. Use logarithms to compute the value of $M = \frac{1}{2}x^2 \cos (35° + \theta)$, given by:
   (a) $x = 47, \theta = 53°$;   (b) $x = 0.537, \theta = 68°$;

(c) $x = 0.0378$, $\theta = 70°30'$;   (d) $x = 568$, $\theta = 0°30'$;

(e) $x = 9.09$, $\theta = 91°$;   (f) $x = 5070$, $\theta = 53°30'$.

4.   Use logarithms to compute:

(a) $\dfrac{135}{\sin 49°}$,

(b) $\dfrac{0.43 \sin 54°}{\sin 27°}$,

(c) $\dfrac{237 \cos 86°}{\tan 45° \cot 40°}$,

(d) $237^3 \sin 18° \cos 35°$,

(e) $\dfrac{\sqrt{314 \times 26.5 \tan 51°}}{\sin 51°}$,

(f) $\dfrac{\sin^3 63° \cos 12°}{4.2 \tan^2 49°}$,   (recall the notation $\sin^3 \alpha = (\sin \alpha)^3$),

(g) $0.0376^2 \dfrac{\sin 37° \sin 86°}{\sin 57°}$,

(h) $\sin 49° \sqrt{\dfrac{3.73}{0.0935}}$,

(i) $45.6 \sqrt{\sin 56° \cos 34°}$,

(j) $\dfrac{3.93}{\sin 67°} + \dfrac{47.6}{\tan 63°}$,

(k) $\sqrt{5.63} + \sqrt[3]{4.82 \cos 98°}$,

(l) $\dfrac{4.07 \sin 57°}{9.9 \cos 36°} + \sqrt{\dfrac{4.07 \sin 57°}{9.9 \cos 36°}}$,

(m) $5.69 \sin^4 38° \cos^3 73°$,

(n) $\sqrt{6.36 \tan^3 56°30' \cos^2 93°}$,

(o) $\sqrt[3]{\dfrac{58.6 \sin^2 87°}{6.93 \cos^4 20°}}$,

(p) $\sqrt[5]{\dfrac{69.3}{609}} \tan^3 99°$,

(r) $\sqrt[3]{0.0325 \sin 47°} - \sqrt{26.3 \cos 47°}$,

(s) $\sqrt{\dfrac{263}{\sin 81°}} \div \sqrt{\dfrac{\cos 81°}{0.235}}$,

(t) $(\sin 57°)^{3.75}$,

(u) $(\sin 57° + \cos 57°)^{3.75}$,

(v) $35.7 \sin 105°$,

(w) $\dfrac{391^5 \cos^3 23°30'}{\sqrt{\tan 45°}}$,

(x) $\sqrt{36.5^2 \cos^2 73° + 62.1^2 \sin^2 73°}$,

(y) $-\dfrac{0.0573 \sin^3 80°}{2.38 \cos^3 73°}$,

(z) $-\dfrac{23.6^4 \sin^2 38°}{0.89^3 \cos^3 125°}$.

5.   Use logarithms to compute the following to four figure accuracy:

(a) $276.3 \cot 46°20'$,

(b) $46.92 \sin^3 25°15'$,

(c) $\sqrt{\sin 15°45' \cos 31°30'}$,

(d) $\cos 126°50' \operatorname{cosec} 217°$,

(e) $\dfrac{0.06335 \tan^2 63°15'}{0.172 \sin 16°20'}$.

3.4   Inverse Trigonometric Functions

   Many equations occurring in mathematics and physics involve trigonometric functions. Only a few special types of such equations can be solved explicitly and exactly. We shall discuss solutions of trigonometric equations that can be reduced to the form

(1) $f(x) = c,$

where f denotes a trigonometric function and c a number. In some cases (1) will take the more convenient form

$$f(x) = f(c).$$

To solve (1) we first find all the solutions in the interval $0° \le x < 360°$. If c is in the range of f, there are one or two solutions which can be found from the tables with the aid of the reduction formulas given in Section 3.2.

Example 4.1   (a)   The equation

$$\sin x = 2$$

has no solution since 2 is not in the range of the sine function, $[-1, 1]$. (Recall that $[a, b]$ denotes the interval $\{x \mid x \in R, a \le x \le b\}$, while (a, b) denotes $\{x \mid x \in R, a < x < b\}$.)

(b)   Find all solutions of

$$\sin x = 1.$$

In the interval $[0°, 360°)$ there is only one solution, $x = 90°$; hence all the solutions are

$$\{x \mid x = 90° + k \cdot 360°, k \in Z\}.$$

(Recall that Z denotes the set of integers.)

(c)   Find all the solutions of

$$\sin x = 0.8090.$$

We find from Table 2 that $\sin 54° = 0.8090$. Therefore $x = 54°$ is a solution of the equation. Now the sine function increases from 0 to 1 in the first quadrant, decreases from 1 to 0 in the second quadrant, and takes negative values in the other two quadrants. Hence, our geometric intuition tells us that in the interval $[0°, 360°)$, the sine function takes any given positive value at most twice. Now

$$\sin (180° - x) = \sin x,$$

and therefore $180° - 54° = 126°$ is the only other solution in $[0°, 360°)$. Thus all the solutions of $\sin x = 0.8090$ are

$$\{x \mid x = 54° + k \cdot 360° \quad or \quad x = 126° + k \cdot 360°, k \in Z\}.$$

(d)   Solve

(2) $$2 \cos^2 x + \sin x = 1.$$

First we substitute $1 - \sin^2 x$ for $\cos^2 x$:

$$2(1 - \sin^2 x) + \sin x = 1,$$

i.e.,

$$2 \sin^2 x - \sin x - 1 = 0,$$

or

$$(2 \sin x + 1)(\sin x - 1) = 0.$$

Now, the solution set of $2 \sin x + 1 = 0$ is the set of all x for which $\sin x = -\frac{1}{2}$, i.e.,

$$\{x \mid x = 210° + k \cdot 360° \quad or \quad x = 330° + k \cdot 360°, \ k \in Z\},$$

and the solution set of $\sin x - 1 = 0$ is

$$\{x \mid x = 90° + k \cdot 360°, \ k \in Z\}.$$

The union of these two sets is the solution set of (2).

(e) Find all solutions of

(3) $\cos x = \cos 37°.$

Clearly $x = 37°$ is a solution. Also $x = -37°$ is a solution. Thus the two solutions in the interval $[0°, 360°)$ are $x = 37°$ and $x = 360° - 37° = 323°$. Hence the solution set of (3) is

$$\{x \mid x = 37° + k \cdot 360° \quad or \quad x = 323° + k \cdot 360°, \ k \in Z\}.$$

We see from the above examples that if we know one solution of

(4) $\sin x = c,$

say $x = \theta$, then the solution set of (4) is

$$\{x \mid x = \theta + k \cdot 360° \quad or \quad x = 180° - \theta + k \cdot 360°, \ k \in Z\}.$$

Similarly, if $x = \theta$ is a solution of

(5) $\cos x = c,$

then the solution set of (5) is

$$\{x \mid x = \pm\theta + k \cdot 360°, \ k \in Z\};$$

and if $x = \theta$ is a solution of

(6) $\tan x = c,$

then the solution set of (6) is

$$\{x \mid x = \theta + k \cdot 180°, \ k \in Z\}.$$

Thus an essential part in solving an equation of the type (4) (or (5) or (6)) is to answer the question: For what angle θ is $\sin \theta = c$ (or $\cos \theta = c$ or $\tan \theta = c$)? In general we are searching for a function related to the sine function the way an inverse function f^{-1} is related to f. We saw in Section 2.1 that the inverse function is defined only if f is 1–1 and onto. The trigonometric functions as defined in Section 3.2 are neither. However, if we restrict the domain of the sine function to the interval $[-\pi/2, \pi/2]$ (or, in degrees, to $[-90°, 90°]$), then the restricted function is one–one and onto $[-1, 1]$. That is to say, if the function $f: [-\pi/2, \pi/2] \rightarrow [-1, 1]$ is defined by

$$f(x) = \sin x,$$

then f is one–one and onto $[-1, 1]$, and therefore $f^{-1}: [-1, 1] \rightarrow [-\pi/2, \pi/2]$ exists. It is customary to write $\sin^{-1}$ instead of f^{-1}.

We can similarly restrict the domain of the cosine function to $[0, \pi]$ and the domain of the tangent function to $(-\pi/2, \pi/2)$, and define the inverse functions

$$\cos^{-1}: [-1, 1] \rightarrow [0, \pi],$$
$$\tan^{-1}: R \rightarrow (-\pi/2, \pi/2).$$

Note that the range of each of the inverse functions is a set of angles. These angles may be expressed in radians, degrees, or in any other measure. Thus, for example, $\sin^{-1}(\frac{1}{2}) = \pi/6$ radians $= 30°$. However, it is often more convenient to express the values of inverse trigonometric functions in radians, and hence we shall usually make this our practice.

It is important to remember that $\sin^{-1}$, $\cos^{-1}$, and $\tan^{-1}$ are not the inverse functions of the sine, the cosine, and the tangent functions, respectively. (Indeed neither sine nor cosine nor tangent is one–one!) In fact, $\sin^{-1}(\sin x)$ in general is not equal to x. For example,

$$\sin^{-1}(\sin 5\pi/6) = \sin^{-1} 1/2$$
$$= \pi/6.$$

This happens because $5\pi/6 \notin [-\pi/2, \pi/2]$. Similarly, if $x \notin [0, \pi]$, then $\cos^{-1}(\cos x) \neq x$, and if $x \notin (-\pi/2, \pi/2)$ then $\tan^{-1}(\tan x) \neq x$. In many cases, however, there is no difficulty in determining the values of a composite of an inverse trigonometric function and a direct trigonometric function. For example, we can use our tables to evaluate

$$\sin^{-1}(\cos 2) = \sin^{-1}(-\cos(\pi - 2))$$
$$= -\sin^{-1}(\cos 1.1416)$$
$$= -\sin^{-1}(0.4147)$$
$$= -0.4276.$$

However, it is more elegant to use some fancy algebraic footwork, if possible, rather than tables. Thus

$$\sin^{-1}(\cos 2) = \sin^{-1}(\sin(\pi/2 - 2))$$
$$= \pi/2 - 2$$

since $\pi/2 - 2 \in [-\pi/2, \pi/2]$. Needless to say, $\pi/2 - 2$ is the exact value of $\sin^{-1}(\cos 2)$ while -0.4276 is only an approximate value.

Example 4.2 (a)
$$\sin^{-1}(\sin 5\pi/4) = \sin^{-1}(-\sin \pi/4)$$
$$= -\sin^{-1}(\sin \pi/4)$$
$$= -\pi/4.$$

(b)
$$\sin(\sin^{-1} 0.32) = 0.32.$$

(c) Evaluate $\sin(\cos^{-1} x)$.

We use some trigonometric identities to solve (c). If $0 \le x \le 1$, then $\cos^{-1} x = \sin^{-1}\sqrt{1 - x^2}$. For, if $\cos t = x$, $0 \le t \le \pi/2$, then $\sin t = \sqrt{1 - \cos^2 t} = \sqrt{1 - x^2}$. Hence, in this case,

$$\sin(\cos^{-1} x) = \sin(\sin^{-1}\sqrt{1 - x^2})$$
$$= \sqrt{1 - x^2}.$$

If $-1 \le x \le 0$, then $\pi \ge \cos^{-1} x \ge \pi/2$, and therefore

$$\cos^{-1} x = \pi - \sin^{-1}\sqrt{1 - x^2}.$$

Thus, in this case, we also have

$$\sin(\cos^{-1} x) = \sin(\pi - \sin^{-1}\sqrt{1 - x^2})$$
$$= \sin(\sin^{-1}\sqrt{1 - x^2})$$
$$= \sqrt{1 - x^2}.$$

SYNOPSIS

The present section began with a discussion of the methods for solving simple equations involving trigonometric functions. We then defined the functions $\sin^{-1}$, $\cos^{-1}$, $\tan^{-1}$ and discussed their relations as the inverses to the sine, cosine, and tangent functions.

QUIZ

Answer *true* or *false:*

1. If $\sin x \neq 0$, $\sin^{-1} x = \dfrac{1}{\sin x}$.

2. $\sin^{-1} (\sin x) = x$, for all angles x.
3. $\sin (\sin^{-1} x) = x$, for all x in the domain of $\sin^{-1}$.
4. $\sin^{-1} (-1/2) = -\pi/6$.
5. $\cos^{-1} (-1/2) = -\pi/3$.
6. $\tan^{-1} (\sin \pi/2) = \pi/4$.
7. $\tan^{-1} (\sin \pi) = \pi$.
8. $\sin \pi = \tan \pi$.
9. $\cos^2 x = \cos x$, for all $x = k \cdot \pi/2$, $k \in Z$.
10. $\tan^{-1} x + \tan^{-1} (-x) = 0$, for all $x \in (-\pi/2, \pi/2)$.

EXERCISES

1. Find the value of each of the following angles in degrees:
    (a) $\sin^{-1} 0$,   (b) $\cos^{-1} 0$,   (c) $\tan^{-1} 0$,
    (d) $\sin^{-1} 1$,   (e) $\cos^{-1} 1$,   (f) $\tan^{-1} 1$,
    (g) $\sin^{-1} (-1)$,   (h) $\cos^{-1} (-1)$,   (i) $\tan^{-1} (-1)$,
    (j) $\sin^{-1} (-\frac{1}{2})$,   (k) $\cos^{-1} \frac{1}{2}$,   (l) $\cos^{-1} (-\frac{1}{2})$,
    (m) $\sin^{-1} (-\sqrt{3}/2)$,   (n) $\tan^{-1} \sqrt{3}$,

    (o) $\tan^{-1} \dfrac{1}{\sqrt{3}}$,   (p) $\tan^{-1} \left(-\dfrac{1}{\sqrt{3}}\right)$,

    (q) $\sin^{-1} \left(-\dfrac{1}{\sqrt{2}}\right)$,   (r) $\cos^{-1} \left(-\dfrac{1}{\sqrt{2}}\right)$,

    (s) $\cos^{-1} (\cos 4\pi/3)$,   (t) $\sin^{-1} (\sin 2\pi/3)$.
2. Find each of the angles in Exercise 1 in radian measure.
3. Use Table 2 to find the value of each of the following angles in degrees:
    (a) $\sin^{-1} 0.5225$,   (b) $\cos^{-1} 0.5225$,
    (c) $\tan^{-1} 0.7400$,   (d) $\sin^{-1} 0.7771$,
    (e) $\cos^{-1} (-0.6495)$,   (f) $\sin^{-1} (-0.6495)$,
    (g) $\tan^{-1} 4.915$,   (h) $\sin^{-1} 0.1908$,
    (i) $\cos^{-1} (-0.1908)$,   (j) $\tan^{-1} (19.08)$.
4. Find the value of each of the following:
    (a) $\sin (\sin^{-1} \frac{1}{2})$,   (b) $\cos (\cos^{-1} \frac{1}{3})$,
    (c) $\tan (\tan^{-1} 2)$,   (d) $\sin^{-1} (\sin 30°)$,
    (e) $\cos^{-1} (\cos 10°)$,   (f) $\tan^{-1} (\tan \pi/3)$,
    (g) $\sin^{-1} (\sin 2\pi/3)$,   (h) $\cos^{-1} (\cos 4\pi/3)$,

(i) $\tan^{-1}(\tan 4\pi/3)$, (j) $\sin^{-1}(\sin(-30°))$,

(k) $\cos^{-1}(\cos(-30°))$.

5. Find the value of each of the following:

(a) $\sin(\cos^{-1}4/5)$, (b) $\cos(\sin^{-1}1/2)$,

(c) $\sin(\tan^{-1}3/4)$, (d) $\sin(\cos^{-1}2/3)$,

(e) $\csc(\sin^{-1}2/5)$, (f) $\cot(\cos^{-1}1/2)$,

(g) $\cot(\tan^{-1}3)$, (h) $\cos(\sin^{-1}(-2/3))$,

(i) $\tan(\sin^{-1}(-2/3))$, (j) $\tan\left(\frac{\pi}{2}-\sin^{-1}1/3\right)$,

(k) $\sin\left(\frac{\pi}{2}+\tan^{-1}3/4\right)$

$$\left[Hint:\ \sin\left(\frac{\pi}{2}+\alpha\right)=\sin\left(\pi-\left(\frac{\pi}{2}+\alpha\right)\right)\right.$$

$$\left.=\sin\left(\frac{\pi}{2}-\alpha\right)=\cos\alpha,\ \ \text{for all}\ \ \alpha\right],$$

(l) $\cos(\pi-\cos^{-1}3/4)$, (m) $\tan(\pi+\sin^{-1}2/3)$.

6. Find the value of each of the following:

(a) $\sin(\cos^{-1}x)$, (b) $\sin(\tan^{-1}x)$,

(c) $\cos(\sin^{-1}x)$, (d) $\cos(\tan^{-1}x)$,

(e) $\tan(\sin^{-1}x)$, (f) $\tan(\cos^{-1}x)$,

(g) $\csc(\sin^{-1}1/x)$, (h) $\sec(\tan^{-1}\sqrt{x^2-1})$, $x>1$.

7. Find all angles x which satisfy the following equations:

(a) $\sin x \cos x = 0$;

(b) $\sin 2x = \cos 2x$;

(c) $3\sin x = 2\cos^2 x$;

(d) $3\cos(x+30°)+2\sin^2(x+30°)=3$;

(e) $\sin x + \cos x = 1$ [*Hint:* Square both sides of $\sin x = 1 - \cos x$ and use Theorem 2.2, formula (29). For alternative methods see Examples 1.2(b) and (d) in the next section.]

8. Suggest appropriate definitions for $\csc^{-1}$, $\sec^{-1}$, and $\cot^{-1}$.

Trigonometric Identities

4.1 Sums of Angles

In this section we shall further develop the trigonometric formulas. This announcement may alarm the reader who is perhaps already apprehensive about the profusion of trigonometric formulas in Section 3.2. Some justification is therefore in order. Trigonometric functions occur in many problems in pure and applied mathematics. The solutions of such problems often involve complicated expressions that are intractable until simplified by the use of trigonometric identities. These identities are also useful for transforming trigonometric equations to forms solvable by elementary algebraic methods. We saw in Section 3.3 that some of the identities from Section 3.2 can be used to simplify the construction of trigonometric tables. Trigonometric identities are also useful in transforming trigonometric formulas into a "multiplicative" form suitable for computing by means of log tables. The basic formula from which all others will be derived is given in the following theorem.

Theorem 1.1 *Let α and β be any angles. Then*

$$\cos (\alpha - \beta) = \cos \alpha \cos \beta + \sin \alpha \sin \beta.$$

Proof Let OA be the initial side and OP, OQ, OR the terminal sides of the angles α, β, $\alpha - \beta$, respectively (Figure 4.1). Since $\alpha - \beta$ is the difference between the rotations α and β, it follows that the length of the arc QP is equal to the length of the arc AR, and therefore the distance QP is equal to the distance AR. Now, the coordinates of points P, Q, R, and A are $(\cos \alpha, \sin \alpha)$, $(\cos \beta, \sin \beta)$, $(\cos (\alpha - \beta), \sin (\alpha - \beta))$, and $(1, 0)$, respectively. Therefore

134

Figure 4.1

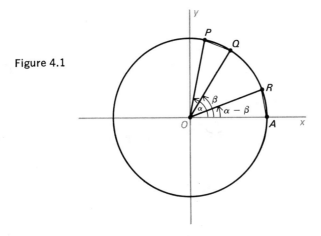

$$QP^2 = (\cos \alpha - \cos \beta)^2 + (\sin \alpha - \sin \beta)^2$$
$$= \cos^2 \alpha - 2 \cos \alpha \cos \beta + \cos^2 \beta + \sin^2 \alpha - 2 \sin \alpha \sin \beta + \sin^2 \beta$$
$$= (\sin^2 \alpha + \cos^2 \alpha) + (\sin^2 \beta + \cos^2 \beta) - 2(\cos \alpha \cos \beta + \sin \alpha \sin \beta)$$
$$= 2 - 2(\cos \alpha \cos \beta + \sin \alpha \sin \beta),$$

and

$$AR^2 = \big(\cos (\alpha - \beta) - 1\big)^2 + \big(\sin (\alpha - \beta) - 0\big)^2$$
$$= \cos^2 (\alpha - \beta) - 2 \cos (\alpha - \beta) + 1 + \sin^2 (\alpha - \beta)$$
$$= 2 - 2 \cos (\alpha - \beta).$$

Since $AR^2 = QP^2$, we have

$$2 - 2 \cos (\alpha - \beta) = 2 - 2(\cos \alpha \cos \beta + \sin \alpha \sin \beta),$$

i.e.,

$$\cos (\alpha - \beta) = \cos \alpha \cos \beta + \sin \alpha \sin \beta. \ \blacksquare$$

Theorem 1.2 *Let α and β be any angles. Then:*

(1) $$\sin (\alpha + \beta) = \sin \alpha \cos \beta + \cos \alpha \sin \beta;$$

(2) $$\sin (\alpha - \beta) = \sin \alpha \cos \beta - \cos \alpha \sin \beta;$$

(3) $$\cos (\alpha + \beta) = \cos \alpha \cos \beta - \sin \alpha \sin \beta;$$

(4) $$\cos (\alpha - \beta) = \cos \alpha \cos \beta + \sin \alpha \sin \beta;$$

(5) $$\tan (\alpha + \beta) = \frac{\tan \alpha + \tan \beta}{1 - \tan \alpha \tan \beta}$$

whenever both sides are defined;

$$(6) \qquad \tan (\alpha - \beta) = \frac{\tan \alpha - \tan \beta}{1 + \tan \alpha \tan \beta}$$

whenever both sides are defined.

Proof Formula (4) was proved in Theorem 1.1. We quote it here for completeness. Hence, using (4) and formulas (8) and (9) in Section 3.2, we have

$$\begin{aligned}
\cos (\alpha + \beta) &= \cos \left(\alpha - (-\beta) \right) \\
&= \cos \alpha \cos (-\beta) + \sin \alpha \sin (-\beta) \\
&= \cos \alpha \cos \beta - \sin \alpha \sin \beta,
\end{aligned}$$

which is (3). Now using (4) and formulas (12) and (13) in Section 3.2:

$$\begin{aligned}
\sin (\alpha + \beta) &= \cos \left(90° - (\alpha + \beta) \right) \\
&= \cos \left((90° - \alpha) - \beta \right) \\
&= \cos (90° - \alpha) \cos \beta + \sin (90° - \alpha) \sin \beta \\
&= \sin \alpha \cos \beta + \cos \alpha \sin \beta,
\end{aligned}$$

which is formula (1). To prove (5) suppose that α and β are any angles such that $\cos (\alpha + \beta) \neq 0$, and therefore $\tan (\alpha + \beta)$ is defined. Then, by (1) and (3),

$$\begin{aligned}
\tan (\alpha + \beta) &= \frac{\sin (\alpha + \beta)}{\cos (\alpha + \beta)} \\
&= \frac{\sin \alpha \cos \beta + \cos \alpha \sin \beta}{\cos \alpha \cos \beta - \sin \alpha \sin \beta}.
\end{aligned}$$

If $\cos \alpha \cos \beta \neq 0$, we can divide the numerator and the denominator of the fraction by $\cos \alpha \cos \beta$ and obtain

$$\begin{aligned}
\tan (\alpha + \beta) &= \frac{\dfrac{\sin \alpha \cos \beta}{\cos \alpha \cos \beta} + \dfrac{\cos \alpha \sin \beta}{\cos \alpha \cos \beta}}{\dfrac{\cos \alpha \cos \beta}{\cos \alpha \cos \beta} - \dfrac{\sin \alpha \sin \beta}{\cos \alpha \cos \beta}} \\
&= \frac{\tan \alpha + \tan \beta}{1 - \tan \alpha \tan \beta}.
\end{aligned}$$

If $\cos \alpha \cos \beta = 0$, then either $\cos \alpha = 0$ or $\cos \beta = 0$, i.e., either α or β (or both) differs from 90° by some multiple of 180°. In any case, $\tan \alpha$ or $\tan \beta$ is then undefined, and thus the right-hand side of (5) is undefined. This proves (5). We leave the proofs of formulas (2) and (6) as exercises for the reader. ∎

Of special interest are the formulas obtained by setting $\alpha = \beta$ in (1), (3), and (5):

(7)
$$\sin 2\alpha = 2 \sin \alpha \cos \alpha,$$

(8)
$$\cos 2\alpha = \cos^2 \alpha - \sin^2 \alpha,$$

(9)
$$\tan 2\alpha = \frac{2 \tan \alpha}{1 - \tan^2 \alpha}.$$

Example 1.1 Show that

(10)
$$\sin^2 \alpha = \tfrac{1}{2}(1 - \cos 2\alpha),$$

(11)
$$\cos^2 \alpha = \tfrac{1}{2}(1 + \cos 2\alpha).$$

By Theorem 2.2 of the preceding chapter,

(12)
$$\cos^2 \alpha = 1 - \sin^2 \alpha$$

and

(13)
$$\sin^2 \alpha = 1 - \cos^2 \alpha.$$

Substituting in turn in formula (8), we get

(14)
$$\begin{aligned}\cos 2\alpha &= \cos^2 \alpha - \sin^2 \alpha \\ &= (1 - \sin^2 \alpha) - \sin^2 \alpha \\ &= 1 - 2 \sin^2 \alpha,\end{aligned}$$

and

(15)
$$\begin{aligned}\cos 2\alpha &= \cos^2 \alpha - \sin^2 \alpha \\ &= \cos^2 \alpha - (1 - \cos^2 \alpha) \\ &= 2 \cos^2 \alpha - 1,\end{aligned}$$

from which formulas (10) and (11) immediately follow.

Many of the preceding identities are useful in adapting trigonometric formulas to numerical computations by means of log tables. The underlying idea is to transform formulas involving sums or differences of sines and cosines into products and thus to reduce the number of steps required in using the log tables.

Theorem 1.3 *If λ and μ are any angles, then*

(16)
$$\sin \lambda + \sin \mu = 2 \sin \frac{\lambda + \mu}{2} \cos \frac{\lambda - \mu}{2},$$

(17) $$\sin \lambda - \sin \mu = 2 \cos \frac{\lambda + \mu}{2} \sin \frac{\lambda - \mu}{2},$$

(18) $$\cos \lambda + \cos \mu = 2 \cos \frac{\lambda + \mu}{2} \cos \frac{\lambda - \mu}{2},$$

(19) $$\cos \lambda - \cos \mu = -2 \sin \frac{\lambda + \mu}{2} \sin \frac{\lambda - \mu}{2}.$$

Proof   Set

$$\alpha = \frac{\lambda + \mu}{2} \quad \text{and} \quad \beta = \frac{\lambda - \mu}{2}$$

in formulas (1) and (2). Then

$$\alpha + \beta = \lambda \quad \text{and} \quad \alpha - \beta = \mu,$$

and the two formulas read

(20) $$\sin \lambda = \sin \frac{\lambda + \mu}{2} \cos \frac{\lambda - \mu}{2} + \cos \frac{\lambda + \mu}{2} \sin \frac{\lambda - \mu}{2},$$

(21) $$\sin \mu = \sin \frac{\lambda + \mu}{2} \cos \frac{\lambda - \mu}{2} - \cos \frac{\lambda + \mu}{2} \sin \frac{\lambda - \mu}{2}.$$

If we add the identities (20) and (21), we obtain (16). Subtracting (21) from (20) yields (17). We leave the proofs of identities (18) and (19) as exercises for the reader. ▌

Example 1.2   (a)   Simplify

$$\left(\frac{\sin 3\theta + \sin \theta}{\sin 3\theta - \sin \theta} \right) \left(\frac{\cos 3\theta + \cos \theta}{\cos 3\theta - \cos \theta} \right).$$

Using the identities in Theorem 1.3 with $\lambda = 3\theta$ and $\mu = \theta$, we have

$$\sin 3\theta + \sin \theta = 2 \sin 2\theta \cos \theta,$$
$$\sin 3\theta - \sin \theta = 2 \sin \theta \cos 2\theta,$$
$$\cos 3\theta + \cos \theta = 2 \cos 2\theta \cos \theta,$$
$$\cos 3\theta - \cos \theta = -2 \sin 2\theta \sin \theta.$$

Hence

$$\left(\frac{\sin 3\theta + \sin \theta}{\sin 3\theta - \sin \theta}\right)\left(\frac{\cos 3\theta + \cos \theta}{\cos 3\theta - \cos \theta}\right) = \frac{2 \sin 2\theta \cos \theta}{2 \sin \theta \cos 2\theta} \frac{2 \cos 2\theta \cos \theta}{(-2 \sin 2\theta \sin \theta)}$$

$$= -\frac{\cos^2 \theta}{\sin^2 \theta}$$

$$= -\cot^2 \theta.$$

(b) Solve the equation

(22) $$\sin \theta + \cos \theta = \frac{1}{\sqrt{2}}$$

in the interval $0° \le \theta < 180°$.

Since $\cos \theta = \sin (90° - \theta)$ the left-hand side of (22) can be transformed using (16):

$$\sin \theta + \cos \theta = \sin \theta + \sin (90° - \theta)$$

$$= 2 \sin \frac{\theta + (90° - \theta)}{2} \cos \frac{\theta - (90° - \theta)}{2}$$

$$= 2 \sin 45° \cos (\theta - 45°)$$

$$= \sqrt{2} \cos (\theta - 45°).$$

Thus (22) becomes

$$\sqrt{2} \cos (\theta - 45°) = \frac{1}{\sqrt{2}},$$

i.e.,

(23) $$\cos (\theta - 45°) = \tfrac{1}{2}.$$

Now θ is to be in the interval $0° \le \theta < 180°$ and thus $-45° \le \theta - 45° < 135°$. There is only one solution of (23), and therefore of (22), in this interval:

$$\theta - 45° = 60°,$$

and hence

$$\theta = 105°.$$

(c) If c, d are given positive numbers and α is a given angle, find a number r and an angle ϵ such that

(24) $$c \sin \alpha + d \cos \alpha = r \sin (\alpha + \epsilon).$$

By (1), the problem is to determine r and ϵ such that

$$c \sin \alpha + d \cos \alpha = r \cos \epsilon \sin \alpha + r \sin \epsilon \cos \alpha.$$

It suffices therefore to find r and ϵ satisfying

(25) $$r \cos \epsilon = c$$

and

(26) $$r \sin \epsilon = d.$$

Squaring both sides of (25) and (26) and adding, we obtain

$$r^2(\sin^2 \epsilon + \cos^2 \epsilon) = c^2 + d^2,$$

i.e.,

$$r^2 = c^2 + d^2$$

by Theorem 2.2, Section 3.2. Set

$$r = \sqrt{c^2 + d^2},$$

and

$$\epsilon = \cos^{-1} \frac{c}{\sqrt{c^2 + d^2}}.$$

Then (25) is satisfied, and

$$
\begin{aligned}
r \sin \epsilon &= \sqrt{c^2 + d^2} \sqrt{1 - \cos^2 \epsilon} \\
&= \sqrt{c^2 + d^2} \sqrt{1 - \frac{c^2}{c^2 + d^2}} \\
&= \sqrt{c^2 + d^2} \sqrt{\frac{d^2}{c^2 + d^2}} \\
&= d.
\end{aligned}
$$

Thus (26) is satisfied as well.

(d)   Use the method in Example (c) to solve equation (22).
Setting $c = d = 1$ and $\alpha = \theta$ in (24), we obtain

$$
\begin{aligned}
r &= \sqrt{1^2 + 1^2} \\
&= \sqrt{2},
\end{aligned}
$$

$$\epsilon = \cos^{-1}\frac{1}{\sqrt{2}}$$

$$= 45°.$$

Therefore

$$\sin\theta + \cos\theta = \sqrt{2}\sin(\theta + 45°).$$

Substituting this in equation (22), we obtain

$$\sqrt{2}\sin(\theta + 45°) = \frac{1}{\sqrt{2}},$$

i.e.,

(27) $$\sin(\theta + 45°) = \tfrac{1}{2}.$$

Now $0° \le \theta < 180°$ and therefore $45° \le \theta + 45° < 225°$. Hence (27) has only one solution

$$\theta + 45° = 150°,$$

or

$$\theta = 105°.$$

SYNOPSIS

In this section we proved important trigonometric identities involving sums and differences of angles. In particular, we obtained formulas for $\sin 2\alpha$, $\cos 2\alpha$, and $\tan 2\alpha$ in terms of $\sin \alpha$, $\cos \alpha$, and $\tan \alpha$. Finally, we established formulas converting a sum of two trigonometric functions into a product of two functions.

QUIZ

Answer *true* or *false:*
1. $\cos^2 45° = 1 - \cos^2 45°$.
2. $\cos 2\alpha = 1 - 2\sin^2 \alpha$, for all α.
3. $\sin 2\alpha = 1 - 2\cos^2 \alpha$, for all α.
4. $\cos \alpha - \cos \beta = 2\sin\frac{\beta + \alpha}{2}\sin\frac{\beta - \alpha}{2}$, for all α and β.
5. If $\sin(\alpha + \beta) = \sin \alpha$ and $0° \le \beta < 360°$, then $\beta = 0°$.
6. $\sin\theta = 2\sin\frac{1}{2}\theta\cos\frac{1}{2}\theta$, for all θ.
7. $(\sin\theta + \cos\theta)^2 = 1 + 2\sin\theta\cos\theta$, for all θ.
8. $\sin 2\theta - \sin\theta = \sin\theta$, for all θ.

9. $\dfrac{\sin 2\theta}{\sin \theta} = \sin \theta$, for all θ.

10. $\dfrac{\sin 2\theta}{2} = \sin \theta$, for all θ.

EXERCISES

1. Prove formula (2).
2. Prove formula (6).
3. Prove formula (18).
4. Prove formula (19).
5. Establish the identities

(28) $$\sin 3\alpha = 3 \sin \alpha - 4 \sin^3 \alpha,$$

(29) $$\cos 3\alpha = 4 \cos^3 \alpha - 3 \cos \alpha.$$

[*Hint:* To prove (28) use (1), (7), (12), and (14); identity (29) is proved similarly.]

6. Use the formulas in Theorem 1.3 to express each of the following as a product:
    (a) $\sin 37° + \sin 43°$,   (b) $\sin 68° + \sin 35°$,
    (c) $\cos 58° + \cos 12°$,   (d) $\cos 39° + \cos 68°$,
    (e) $\sin 27° - \sin 13°$,   (f) $\sin 80° - \sin 47°$,
    (g) $\cos 36° - \cos 82°$,   (h) $\cos 81° - \cos 40°$,
    (i) $\cos 66° + \sin 57°$ [*Hint:* $\cos 66° = \sin (90° - 66°) = \sin 24°$.],
    (j) $\sin 81° - \cos 26°$,   (k) $\cos 12° - \sin 36°$,
    (l) $\sin 5\alpha - \sin 3\alpha$,   (m) $\cos 4\alpha - \cos \alpha$.
7. Prove the following formulas:
    (a) $\sin \alpha \cos \beta = \frac{1}{2}(\sin (\alpha + \beta) + \sin (\alpha - \beta))$;
    (b) $\cos \alpha \sin \beta = \frac{1}{2}(\sin (\alpha + \beta) - \sin (\alpha - \beta))$;
    (c) $\cos \alpha \cos \beta = \frac{1}{2}(\cos (\alpha + \beta) + \cos (\alpha - \beta))$;
    (d) $\sin \alpha \sin \beta = \frac{1}{2}(\cos (\alpha - \beta) - \cos (\alpha + \beta))$.
8. Use the formulas in the preceding exercise to express the following as a sum or a difference of sines or cosines:
    (a) $\sin 67° \cos 21°$,   (b) $\sin 13° \cos 29°$,
    (c) $\sin 26° \sin 54°$,   (d) $\cos 36° \cos 40°$,
    (e) $\sin 80° \sin 70°$,   (f) $\cos 55° \cos 123°$,
    (g) $6 \sin 6x \sin x$,   (h) $4 \cos 4x \cos 3x$,
    (i) $2 \cos 3x \cos 2x$,   (j) $8 \cos 2\alpha \sin \alpha$.
9. Solve the following equations for x in the interval $0° \le x < 360°$:
    (a) $2 \sin x \cos x = 1$;

(b) $6 \sin^2 x + \cos 2x = 2$;

(c) $\cos 2x - \sin x = 1$;

(d) $\tan x + 2 \cos x = \sec x$;

(e) $\sin 5x + \sin 3x = 0$ [*Hint:* Use (16) and recall that a product is zero if and only if at least one of the factors is zero.];

(f) $\cos 2x + \cos 3x = 0$;

(g) $\sin 2x + \sin 4x = \cos x$;

(h) $\cos x + \cos 3x = 2 \sin 2x$;

(i) $\sin x + \sqrt{3} \cos x = 1$ [*Hint:* Use the method in Example 1.2(c).];

(j) $5 \sin x + 12 \cos x = 2$;

(k) $3 \cos x - 4 \sin x = 1$ [*Hint:* Follow the method in Example 1.2(c) using formula (2) in Theorem 1.2 instead of formula (1).];

(l) $\sin x - \sqrt{3} \cos x = 1$.

10. (a) Express $\cos^4 \alpha$ in the form $a_0 + a_1 \cos \alpha + a_2 \cos 2\alpha + a_3 \cos 3\alpha + a_4 \cos 4\alpha$ where a_0, a_1, a_2, a_3, a_4 are fixed numbers independent of α. [*Hint:* Use formula (11).]

(b) Express $\sin^4 \alpha$ in the form $b_0 + b_1 \cos \alpha + b_2 \cos 2\alpha + b_3 \cos 3\alpha + b_4 \cos 4\alpha$ where b_0, b_1, b_2, b_3, b_4 are fixed numbers.

11. Use the data in Table (10), Section 3.1, to compute the exact values (in terms of radicals) of $\sin 15°$, $\cos 15°$, $\tan 15°$, $\sin 75°$, $\cos 75°$, $\tan 75°$.

12. Prove the identities

$$(30) \qquad \sin \alpha = \frac{\tan \alpha}{\sqrt{1 + \tan^2 \alpha}},$$

$$(31) \qquad \cos \alpha = \frac{1}{\sqrt{1 + \tan^2 \alpha}},$$

for $0° \le \alpha \le 90°$.

[*Hint:* Use formula (30) in Section 3.2.]

4.2 Trigonometric Identities

In the preceding sections we obtained many basic identities between various trigonometric functions. These fundamental identities, which we list again below, are used to establish many identities that are of importance in mathematics, physics, and engineering. The purpose of establishing all these identities is to simplify complicated expressions involving trigonometric functions, or to render them more tractable in specific problems such as computations or solution of trigonometric equations.   For example, in the preceding section we solved equation (22) by transforming $\sin \theta + \cos \theta$ into $\sqrt{2} \cos (\theta - 45°)$.

All the identities in the following theorem were previously established in Sections 3.2 and 4.1.   We list them here for reference.

Theorem 2.1   *If α and β are any angles, then the following identities hold whenever the trigonometric functions involved are defined:*

(1) $$\tan \alpha = \frac{\sin \alpha}{\cos \alpha};$$

(2) $$\cot \alpha = \frac{1}{\tan \alpha};$$

(3) $$\operatorname{cosec} \alpha = \frac{1}{\sin \alpha};$$

(4) $$\sec \alpha = \frac{1}{\cos \alpha};$$

(5) $$\sin (-\alpha) = -\sin \alpha;$$

(6) $$\cos (-\alpha) = \cos \alpha;$$

(7) $$\tan (-\alpha) = -\tan \alpha;$$

(8) $$\sin (180° - \alpha) = \sin \alpha;$$

(9) $$\cos (180° - \alpha) = -\cos \alpha;$$

(10) $$\tan (180° - \alpha) = -\tan \alpha;$$

(11) $$\sin (90° - \alpha) = \cos \alpha;$$

(12) $$\cos (90° - \alpha) = \sin \alpha;$$

(13) $$\tan (90° - \alpha) = \cot \alpha;$$

(14) $$\sin^2 \alpha + \cos^2 \alpha = 1;$$

(15) $$\tan^2 \alpha + 1 = \sec^2 \alpha;$$

(16) $$\sin (\alpha + \beta) = \sin \alpha \cos \beta + \cos \alpha \sin \beta;$$

(17) $\qquad \sin(\alpha - \beta) = \sin\alpha\cos\beta - \cos\alpha\sin\beta;$

(18) $\qquad \cos(\alpha + \beta) = \cos\alpha\cos\beta - \sin\alpha\sin\beta;$

(19) $\qquad \cos(\alpha - \beta) = \cos\alpha\cos\beta + \sin\alpha\sin\beta;$

(20) $\qquad \tan(\alpha + \beta) = \dfrac{\tan\alpha + \tan\beta}{1 - \tan\alpha\tan\beta};$

(21) $\qquad \tan(\alpha - \beta) = \dfrac{\tan\alpha - \tan\beta}{1 + \tan\alpha\tan\beta};$

(22) $\qquad \sin 2\alpha = 2\sin\alpha\cos\alpha;$

(23) $\qquad \cos 2\alpha = \cos^2\alpha - \sin^2\alpha;$

(24) $\qquad \tan 2\alpha = \dfrac{2\tan\alpha}{1 - \tan^2\alpha};$

(25) $\qquad \sin^2\alpha = \tfrac{1}{2}(1 - \cos 2\alpha);$

(26) $\qquad \cos^2\alpha = \tfrac{1}{2}(1 + \cos 2\alpha);$

(27) $\qquad \sin\alpha + \sin\beta = 2\sin\dfrac{\alpha + \beta}{2}\cos\dfrac{\alpha - \beta}{2};$

(28) $\qquad \sin\alpha - \sin\beta = 2\sin\dfrac{\alpha - \beta}{2}\cos\dfrac{\alpha + \beta}{2};$

(29) $\qquad \cos\alpha + \cos\beta = 2\cos\dfrac{\alpha + \beta}{2}\cos\dfrac{\alpha - \beta}{2};$

(30) $\qquad \cos\alpha - \cos\beta = -2\sin\dfrac{\alpha + \beta}{2}\sin\dfrac{\alpha - \beta}{2}.$

In order to establish an identity we have to prove that the equality holds true for all angles for which both sides are defined. For example, in order to show that

(31) $\qquad \tan\alpha + \cot\alpha = 1/(\sin\alpha\cos\alpha),$

we start with the left-hand side of (31) and use (1), (2), and (14):

$$\tan\alpha + \cot\alpha = \frac{\sin\alpha}{\cos\alpha} + \frac{\cos\alpha}{\sin\alpha}$$

$$= \frac{\sin^2 \alpha + \cos^2 \alpha}{\sin \alpha \cos \alpha}$$

$$= \frac{1}{\sin \alpha \cos \alpha}.$$

In general, this is the best procedure to prove identities: Start with one side of the identity to be proved and transform it, using identities in Theorem 2.1, into the other side.

Example 2.1   (a)   Prove the identity

(32) $$\sec \alpha - \tan \alpha = \frac{1}{\sec \alpha + \tan \alpha}.$$

Denote by LHS and RHS the left-hand side and the right-hand side of (32), respectively. Then, using (15),

$$\text{RHS} = \frac{1}{\sec \alpha + \tan \alpha}$$

$$= \frac{\sec \alpha - \tan \alpha}{(\sec \alpha + \tan \alpha)(\sec \alpha - \tan \alpha)}$$

$$= \frac{\sec \alpha - \tan \alpha}{\sec^2 \alpha - \tan^2 \alpha}$$

$$= \frac{\sec \alpha - \tan \alpha}{(\tan^2 \alpha + 1) - \tan^2 \alpha}$$

$$= \sec \alpha - \tan \alpha$$

$$= \text{LHS.}$$

The student who has difficulty in memorizing trigonometric formulas involving all six trigonometric functions and in applying these formulas to prove identities, such as (32), can use a less elegant but usually quite efficient method: Reduce each side to its simplest form in terms of sines and cosines. In our present example

$$\text{LHS} = \sec \alpha - \tan \alpha$$

$$= \frac{1}{\cos \alpha} - \frac{\sin \alpha}{\cos \alpha}$$

$$= \frac{1 - \sin \alpha}{\cos \alpha},$$

while the right-hand can be transformed as follows:

$$\text{RHS} = \frac{1}{\sec\alpha + \tan\alpha}$$

$$= \frac{1}{\dfrac{1}{\cos\alpha} + \dfrac{\sin\alpha}{\cos\alpha}}$$

$$= \frac{\cos\alpha}{1 + \sin\alpha}$$

$$= \frac{\cos\alpha(1 - \sin\alpha)}{(1 + \sin\alpha)(1 - \sin\alpha)}$$

$$= \frac{\cos\alpha(1 - \sin\alpha)}{1 - \sin^2\alpha}$$

$$= \frac{\cos\alpha(1 - \sin\alpha)}{\cos^2\alpha}$$

$$= \frac{1 - \sin\alpha}{\cos\alpha}.$$

Thus each side of (32) is equal to $\dfrac{1 - \sin\alpha}{\cos\alpha}$, and hence (32) is proved.

(b) Establish the identity

(33) $$\frac{\sin 2\theta}{\sin\theta} - \sec\theta = \frac{\cos 2\theta}{\cos\theta}.$$

Using (22)

$$\text{LHS} = \frac{2\sin\theta\cos\theta}{\sin\theta} - \frac{1}{\cos\theta}$$

$$= 2\cos\theta - \frac{1}{\cos\theta},$$

while by (26)

$$\text{RHS} = \frac{\cos 2\theta}{\cos\theta}$$

$$= \frac{2\cos^2\theta - 1}{\cos\theta}$$

$$= 2\cos\theta - \frac{1}{\cos\theta}$$

$$= \text{LHS}.$$

In the remainder of this section we show how to prove various kinds of identities.

Example 2.2 Prove the following identities:

(a) $\quad 1 - \dfrac{\cos^2 \alpha}{1 + \sin \alpha} = \sin \alpha;$

(b) $\quad \sin \beta + \cos \beta = \dfrac{1 + 2 \sin \beta \cos \beta}{\sin \beta + \cos \beta};$

(c) $\quad \dfrac{\cot \alpha + \cos \alpha}{\cot \alpha \cos \alpha} = \dfrac{\cot \alpha \cos \alpha}{\cot \alpha - \cos \alpha};$

(d) $\quad \dfrac{\sin 3\alpha}{\sin \alpha} - \dfrac{\cos 3\alpha}{\cos \alpha} = 2;$

(e) $\quad \dfrac{1 + \sin \theta}{\cos \theta} = \tan \left(\dfrac{\theta}{2} + 45° \right);$

(f) $\quad \dfrac{\sin 3\alpha}{\sin \alpha} = 2 \cos 2\alpha + 1;$

(g) $\quad \dfrac{\sin \lambda + \sin 4\lambda + \sin 7\lambda}{\cos \lambda + \cos 4\lambda + \cos 7\lambda} = \tan 4\lambda;$

(h) $\quad \dfrac{3 \sin \beta - \sin 3\beta}{3 \cos \beta + \cos 3\beta} = \tan^3 \beta;$

(i) $\quad \dfrac{\sin 7t}{\operatorname{cosec} t} + \dfrac{\cos 7t}{\sec t} = \cos 6t.$

(a) When one side of the identity to be proved is more complicated than the other, it is preferable to start with that side:

$$
\begin{aligned}
\text{LHS} &= 1 - \frac{\cos^2 \alpha}{1 + \sin \alpha} \\
&= 1 - \frac{1 - \sin^2 \alpha}{1 + \sin \alpha} \\
&= 1 - \frac{(1 - \sin \alpha)(1 + \sin \alpha)}{1 + \sin \alpha} \\
&= 1 - (1 - \sin \alpha) \\
&= \sin \alpha \\
&= \text{RHS.}
\end{aligned}
$$

(b) Both sides of this identity are expressions involving only $\sin \beta$ and $\cos \beta$ and each side is apparently in simplest form. We try to transform the LHS into a form resembling the RHS; we write it as a fraction with the same denominator as the fraction on the RHS:

$$\text{LHS} = \sin \beta + \cos \beta$$

$$= \frac{(\sin \beta + \cos \beta)^2}{\sin \beta + \cos \beta}$$

$$= \frac{\sin^2 \beta + 2 \sin \beta \cos \beta + \cos^2 \beta}{\sin \beta + \cos \beta}$$

$$= \frac{(\sin^2 \beta + \cos^2 \beta) + 2 \sin \beta \cos \beta}{\sin \beta + \cos \beta}$$

$$= \frac{1 + 2 \sin \beta \cos \beta}{\sin \beta + \cos \beta}$$

$$= \text{RHS.}$$

(c) Here we can write both sides in terms of sines and cosines and simplify:

$$\text{LHS} = \frac{\cot \alpha + \cos \alpha}{\cot \alpha \cos \alpha}$$

$$= \frac{\dfrac{\cos \alpha}{\sin \alpha} + \cos \alpha}{\dfrac{\cos \alpha}{\sin \alpha} \cos \alpha}$$

$$= \frac{\cos \alpha + \sin \alpha \cos \alpha}{\cos^2 \alpha}$$

$$= \frac{1 + \sin \alpha}{\cos \alpha},$$

while

$$\text{RHS} = \frac{\cot \alpha \cos \alpha}{\cot \alpha - \cos \alpha}$$

$$= \frac{\dfrac{\cos \alpha}{\sin \alpha} \cos \alpha}{\dfrac{\cos \alpha}{\sin \alpha} - \cos \alpha}$$

$$= \frac{\cos^2 \alpha}{\cos \alpha - \sin \alpha \cos \alpha}$$

$$= \frac{1 - \sin^2 \alpha}{(1 - \sin \alpha) \cos \alpha}$$

$$= \frac{(1 - \sin \alpha)(1 + \sin \alpha)}{(1 - \sin \alpha) \cos \alpha}$$

$$= \frac{1 + \sin \alpha}{\cos \alpha}$$

$$= \text{LHS.}$$

Alternatively, we can try to transform the left-hand side directly into the right-hand side:

$$\text{LHS} = \frac{\cot \alpha + \cos \alpha}{\cot \alpha \cos \alpha}$$

$$= \frac{\cot \alpha \cos \alpha}{\cot \alpha - \cos \alpha} \cdot \frac{\cot \alpha - \cos \alpha}{\cot \alpha \cos \alpha} \cdot \frac{\cot \alpha + \cos \alpha}{\cot \alpha \cos \alpha},$$

since the product of the first two fractions is 1. Thus

$$\text{LHS} = \frac{\cot \alpha \cos \alpha}{\cot \alpha - \cos \alpha} \cdot \frac{\cot^2 \alpha - \cos^2 \alpha}{\cot^2 \alpha \cos^2 \alpha}$$

$$= \frac{\cot \alpha \cos \alpha}{\cot \alpha - \cos \alpha} \left(\frac{1}{\cos^2 \alpha} - \frac{1}{\cot^2 \alpha} \right)$$

$$= \frac{\cot \alpha \cos \alpha}{\cot \alpha - \cos \alpha} \left(\frac{1}{\cos^2 \alpha} - \frac{\sin^2 \alpha}{\cos^2 \alpha} \right)$$

$$= \frac{\cot \alpha \cos \alpha}{\cot \alpha - \cos \alpha} \cdot \frac{1 - \sin^2 \alpha}{\cos^2 \alpha}$$

$$= \frac{\cot \alpha \cos \alpha}{\cot \alpha - \cos \alpha} \cdot 1$$

$$= \text{RHS}.$$

(d)
$$\text{LHS} = \frac{\sin 3\alpha}{\sin \alpha} - \frac{\cos 3\alpha}{\cos \alpha}$$

$$= \frac{\sin 3\alpha \cos \alpha - \cos 3\alpha \sin \alpha}{\sin \alpha \cos \alpha}$$

$$= \frac{\sin (3\alpha - \alpha)}{\sin \alpha \cos \alpha},$$

by (17), and thus using (22),

$$\text{LHS} = \frac{\sin 2\alpha}{\sin \alpha \cos \alpha}$$

$$= \frac{2 \sin \alpha \cos \alpha}{\sin \alpha \cos \alpha}$$

$$= 2$$

$$= \text{RHS}.$$

(e)
$$\text{RHS} = \tan \left(\frac{\theta}{2} + 45° \right)$$

$$= \frac{\tan \frac{\theta}{2} + \tan 45°}{1 - \tan \frac{\theta}{2} \tan 45°},$$

$$= \frac{\tan \frac{\theta}{2} + 1}{1 - \tan \frac{\theta}{2}},$$

since $\tan 45° = 1$. Multiply the numerator and the denominator of the last fraction by $\cos \frac{\theta}{2}$ (which cannot be 0 if the RHS is defined) to obtain

$$\text{RHS} = \frac{\sin \frac{\theta}{2} + \cos \frac{\theta}{2}}{\cos \frac{\theta}{2} - \sin \frac{\theta}{2}},$$

as $\tan \frac{\theta}{2} \cos \frac{\theta}{2} = \sin \frac{\theta}{2}$ by (1). Now on the left-hand side there appear functions of θ, twice $\frac{\theta}{2}$. Hence by (23),

$$\text{RHS} = \frac{\sin \frac{\theta}{2} + \cos \frac{\theta}{2}}{\cos \frac{\theta}{2} - \sin \frac{\theta}{2}} \frac{\cos \frac{\theta}{2} + \sin \frac{\theta}{2}}{\cos \frac{\theta}{2} + \sin \frac{\theta}{2}}$$

$$= \frac{\left(\sin \frac{\theta}{2} + \cos \frac{\theta}{2} \right)^2}{\cos^2 \frac{\theta}{2} - \sin^2 \frac{\theta}{2}}$$

$$= \frac{\sin^2 \frac{\theta}{2} + 2 \sin \frac{\theta}{2} \cos \frac{\theta}{2} + \cos^2 \frac{\theta}{2}}{\cos \theta}$$

$$= \frac{\left(\sin^2 \frac{\theta}{2} + \cos^2 \frac{\theta}{2} \right) + 2 \sin \frac{\theta}{2} \cos \frac{\theta}{2}}{\cos \theta}$$

$$= \frac{1 + \sin \theta}{\cos \theta}$$

$$= \text{LHS}.$$

(f)

$$\text{LHS} = \frac{\sin 3\alpha}{\sin \alpha}$$

$$= \frac{\sin 2\alpha \cos \alpha + \cos 2\alpha \sin \alpha}{\sin \alpha}$$

$$= \frac{\sin 2\alpha \cos \alpha}{\sin \alpha} + \cos 2\alpha$$

$$= \frac{2 \sin \alpha \cos \alpha \cos \alpha}{\sin \alpha} + \cos 2\alpha$$

$$= 2 \cos^2 \alpha + \cos 2\alpha$$

$$= (1 + \cos 2\alpha) + \cos 2\alpha$$

$$= 2 \cos 2\alpha + 1$$

$$= \text{RHS},$$

since $2 \cos^2 \alpha = 1 + \cos 2\alpha$, by (26).

(g)

$$\text{LHS} = \frac{\sin 7\lambda + \sin \lambda + \sin 4\lambda}{\cos 7\lambda + \cos \lambda + \cos 4\lambda}$$

$$= \frac{2 \sin \dfrac{7\lambda + \lambda}{2} \cos \dfrac{7\lambda - \lambda}{2} + \sin 4\lambda}{2 \cos \dfrac{7\lambda + \lambda}{2} \cos \dfrac{7\lambda - \lambda}{2} + \cos 4\lambda}$$

$$= \frac{2 \sin 4\lambda \cos 3\lambda + \sin 4\lambda}{2 \cos 4\lambda \cos 3\lambda + \cos 4\lambda}$$

$$= \frac{(2 \cos 3\lambda + 1) \sin 4\lambda}{(2 \cos 3\lambda + 1) \cos 4\lambda}$$

$$= \frac{\sin 4\lambda}{\cos 4\lambda}$$

$$= \tan 4\lambda$$

$$= \text{RHS}.$$

(h)

$$\text{LHS} = \frac{2 \sin \beta - (\sin 3\beta - \sin \beta)}{2 \cos \beta + (\cos \beta + \cos 3\beta)}$$

$$= \frac{2 \sin \beta - 2 \sin \dfrac{3\beta - \beta}{2} \cos \dfrac{3\beta + \beta}{2}}{2 \cos \beta + 2 \cos \dfrac{3\beta - \beta}{2} \cos \dfrac{3\beta + \beta}{2}}$$

$$= \frac{\sin \beta - \sin \beta \cos 2\beta}{\cos \beta + \cos \beta \cos 2\beta}$$

$$= \frac{\sin \beta \, (1 - \cos 2\beta)}{\cos \beta \, (1 + \cos 2\beta)}$$

$$= \frac{\sin \beta \cdot 2 \sin^2 \beta}{\cos \beta \cdot 2 \cos^2 \beta},$$

by (25) and (26). Thus

$$\text{LHS} = \frac{\sin^3 \beta}{\cos^3 \beta}$$

$$= \tan^3 \beta$$

$$= \text{RHS.}$$

(i)
$$\text{LHS} = \frac{\cos 7t}{\sec t} + \frac{\sin 7t}{\csc t}$$

$$= \cos 7t \cos t + \sin 7t \sin t$$

$$= \cos (7t - t)$$

$$= \cos 6t$$

$$= \text{RHS.}$$

Here we used identities (4), (3), and (19).

SYNOPSIS

Trigonometric identities are of great importance in mathematics, physics, and applied sciences. We listed in this section standard trigonometric identities established in previous sections and gave many worked examples showing the various methods of proving such identities.

QUIZ

Answer *true* or *false:*

1. $\cos \alpha = 2 \cos^2 \frac{\alpha}{2} - 1$, for all α.

2. $\sin \alpha = 2 \sin^2 \frac{\alpha}{2} - 1$, for all α.

3. $\sin 180° \tan (90° - \theta) = \cos 90° \cot (180° - \theta)$ is an identity.

4. $\sin \alpha = 2 \sin \frac{\alpha}{2} \cos \frac{\alpha}{2}$, for all α.

5. $\sin 4\alpha = 2 \sin 2\alpha \cos 2\alpha$, for all α.

6. $\cot \theta = \frac{\sec \theta}{\csc \theta}$, for all θ for which both sides are defined.

7. $\cot \theta = \dfrac{1}{\cot (90° - \theta)}$ is an identity.
8. $\sec^2 \alpha + 1 = \tan^2 \alpha$ is an identity.
9. $2 \sin \lambda \sin (90° - \lambda) = \cos 2(90° - \lambda)$, for all λ.
10. $\operatorname{cosec}^2 \alpha - 1 = \cot^2 \alpha$ is an identity.

EXERCISES

1. Determine which of the following is an identity (i.e., whether the equality holds for all angles for which both sides are defined):
    (a) $\cos \alpha = \cos (360° - \alpha)$;
    (b) $\cos^2 \alpha - \sin^2 (-\alpha) = 1$;
    (c) $\dfrac{\sin 2\alpha}{\sin 4\alpha} = \dfrac{1}{\cos 2\alpha}$;
    (d) $\cos 2\alpha = 2 \sin (45° - \alpha) \cos (\alpha - 45°)$;
    (e) $\dfrac{\sin 3\alpha}{\cos 3\alpha} = \tan \alpha$;
    (f) $\sin 4\theta - \sin 3\theta = \sin \theta$;
    (g) $4 \sin \theta - 3 \sin \theta \overset{.}{=} \sin \theta$;
    (h) $\sin 4\alpha \cos \alpha - \cos 4\alpha \sin \alpha = \sin 3\alpha$;
    (i) $\cos 4\alpha \cos \alpha - \sin 4\alpha \sin \alpha = \cos 3\alpha$;
    (j) $2 \cos^2 \dfrac{\theta}{2} - 1 = \cos \theta$;
    (k) $2 \sin^2 \dfrac{\theta}{2} - 1 = \cos \theta$.

2. Prove the following identities:
    (a) $\dfrac{1 - \cos 2\theta}{\sin 2\theta} = \tan \theta$;
    (b) $\sin 4\alpha = 4 \sin \alpha \cos^3 \alpha - 4 \sin^3 \alpha \cos \alpha$;
    (c) $\sin 2\alpha = 2 \sin \alpha \cos^3 \alpha + 2 \sin^3 \alpha \cos \alpha$;
    (d) $\dfrac{1 + \tan \alpha}{1 - \tan \alpha} = \dfrac{\cos \alpha + \sin \alpha}{\cos \alpha - \sin \alpha}$;
    (e) $\sin^4 \alpha + \cos^4 \alpha = 1 - 2 \sin^2 \alpha \cos^2 \alpha$;
    (f) $\sin^4 \alpha - \cos^4 \alpha = 2 \sin^2 \alpha - 1$;
    (g) $\tan^2 \alpha + \cot^2 \alpha = \operatorname{cosec}^2 \alpha \sec^2 \alpha - 2$;
    (h) $\sin (45° + \alpha) \sec (45° - \alpha) = \cos (45° + \alpha) \operatorname{cosec} (45° - \alpha)$;
    (i) $\dfrac{\cos^2 \alpha}{1 + \sin \alpha} + \dfrac{\cos^2 \alpha}{1 - \sin \alpha} = 2$;
    (j) $\dfrac{1 - \cos \alpha}{1 + \cos \alpha} = (\operatorname{cosec} \alpha - \cot \alpha)^2$;
    (k) $\dfrac{1}{\sec \alpha + \tan \alpha} = \sec \alpha - \tan \alpha$.

3. Prove the following identities:

(a) $\dfrac{2 \tan \alpha}{1 + \tan^2 \alpha} = \sin 2\alpha;$

(b) $\dfrac{\cos 2\theta - \sin \theta}{\cos^2 \theta} = \dfrac{1 - 2 \sin \theta}{1 - \sin \theta}$ [*Hint:* Multiply numerator and denominator of the RHS by $1 + \sin \theta$, and use (25) and (14).];

(c) $\dfrac{1 + \cos 2\lambda}{\sin 2\lambda} = \cot \lambda;$

(d) $\dfrac{\sin \alpha + \sin \beta}{\sin \alpha - \sin \beta} = \dfrac{\tan \frac{1}{2}(\alpha + \beta)}{\tan \frac{1}{2}(\alpha - \beta)};$

(e) $\tan (45° + \alpha) - \tan (45° - \alpha) = 2 \tan 2\alpha;$

(f) $\dfrac{1 + \tan^2 \alpha}{2 \tan \alpha} = \operatorname{cosec} 2\alpha;$

(g) $\dfrac{\cos 2\beta - \sin 2\beta}{\sin \beta \cos \beta} = \cot \beta - \tan \beta - 2;$

(h) $\cos^2 3\theta - \cos^2 \theta = -\sin 2\theta \sin 4\theta;$

(i) $\dfrac{2 \cot \alpha}{1 + \cot^2 \alpha} = \sin 2\alpha;$

(j) $\dfrac{1 + \cos \theta}{1 - \sin \frac{1}{2}\theta} = 2(1 + \sin \frac{1}{2}\theta);$

(k) $\dfrac{\sin 2\alpha \tan 2\alpha}{\tan \alpha} = \sin 2\alpha + \tan 2\alpha;$

(l) $\sin 6x \sin x + \cos 4x \cos 3x = \cos 3x \cos 2x;$

(m) $\dfrac{\sin 10\alpha - \sin 4\alpha}{\sin 4\alpha + \sin 2\alpha} = \dfrac{\cos 7\alpha}{\cos \alpha};$

(n) $\sin 2\theta + \sin 4\theta + \sin 6\theta = 4 \cos \theta \cos 2\theta \sin 3\theta;$

(o) $\dfrac{1 + \cos \theta - 2 \cos \theta \cos \frac{\theta}{2}}{2 - 2 \cos \frac{\theta}{2}} = 1 + \cos \frac{\theta}{2} + \cos \theta$ $\left[Hint: \text{Write the} \right.$

RHS in the form of a fraction with denominator $2 - 2 \cos \dfrac{\theta}{2} \cdot \Big];$

(p) $\dfrac{\sin \theta - \sin 3\theta + \sin 2\theta}{2 - 2 \cos \theta} = \sin \theta + \sin 2\theta.$

Solution of Triangles

5.1 Right-Angled Triangles

In the opening paragraph of Chapter 3 we claimed that trigonometry allows us to "solve" triangles, i.e., to determine the unknown parts of a triangle, given one side and two other parts. In Section 3.1 we indicated how to solve a right-angled triangle. In the present section we review the methods of solution of a right-angled triangle and solve some examples using logarithms.

Recall that if ABC is a triangle right-angled at C (see Figure 3.4), then

(1) $$\sin A = a/c,$$

(2) $$\cos A = b/c,$$

(3) $$\tan A = a/b,$$

where A denotes the angle $\angle BAC$ and a, b, c denote the lengths of BC, CA, AB, respectively. Clearly,

(4) $$\cot A = b/a,$$

(5) $$\operatorname{cosec} A = c/a,$$

(6) $$\sec A = c/b.$$

Example 1.1 (a) Let the triangle ABC be right-angled at C; let $b = 3$ and $c = 4$. Solve the triangle, that is, compute the other side and angles of the triangle. (Again, we are following the convenient custom of denoting an angle in a triangle by the same letter as its vertex. Thus C denotes $\angle BCA$ and A denotes $\angle BAC$.)

Using (2), we have that

$$\cos A = b/c$$
$$= 0.75,$$

and we find from the tables that

$$A = 41°30'.$$

(In order to simplify computations, we shall not use linear interpolation here nor in the remainder of this chapter: we shall evaluate angles to the nearest half degree and lengths to three significant figures.) Now, in a right triangle the sum of the acute angles is 90°, and therefore

$$B = 90° - A$$
$$= 48°30'.$$

The length of the side a can be computed from the Pythagorean theorem or from (1):

$$a = c \sin A$$
$$= 4 \times 0.6626$$
$$= 2.65.$$

(b) See Figure 5.1. In a triangle PQR, the altitude is $PD = 5.24$. The line PD divides $\angle QPR$ internally into $\angle QPD = 25°$ and $\angle RPD = 36°30'$. Find the lengths PQ, PR, and QR.

Figure 5.1

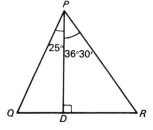

In the triangle QPD we have from (3):

$$QD = PD \tan (\angle QPD)$$
$$= 5.24 \tan 25°.$$

Similarly, in the triangle RDP:

$$DR = 5.24 \tan 36°30'.$$

Numbers	Logs	Antilogs
5.24	0.7193	
$\times$ tan 25°	$+ \bar{1}.6687$	
	0.3880	$2.44 = QD$
5.24	0.7193	
$\times$ tan 36°30'	$+ \bar{1}.8692$	
	0.5885	$3.88 = DR$

Hence,

$$\begin{aligned} QR &= QD + DR \\ &= 2.44 + 3.88 \\ &= 6.32. \end{aligned}$$

In the same two triangles we have, by (2):

$$PQ = \frac{PD}{\cos (\angle QPD)}$$

$$= \frac{5.24}{\cos 25°} \, ;$$

$$PR = \frac{PD}{\cos (\angle RPD)}$$

$$= \frac{5.24}{\cos 36°30'} \, .$$

Numbers	Logs	Antilogs
5.24	0.7193	
$\div$ cos 25°	$- \bar{1}.9573$	
	0.7620	$5.78 = PQ$
5.24	0.7193	
$\div$ cos 36°30'	$- \bar{1}.9052$	
	0.8141	$6.52 = PR$

(c) How long is the side of a regular octagon inscribed in a circle of radius 6.72 inches?

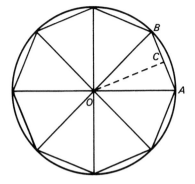

Figure 5.2

Let O be the center of the circle and let AB be a side of the octagon. Let OC be the altitude of the isosceles triangle AOB. Then OC is perpendicular to the base AB and

$$\angle AOB = \frac{360°}{8} = 45°.$$

We compute the length of a side of the octagon:

$$
\begin{aligned}
AB &= 2AC \\
&= 2OA \sin (\angle AOC) \\
&= 2OA \sin \tfrac{1}{2}(\angle AOB) \\
&= 2 \times 6.72 \sin 22°30'.
\end{aligned}
$$

Numbers	Logs	Antilogs
2	0.3010	
× 6.72	+ 0.8274	
× sin 22°30′	+ $\overline{1}$.5828	
	0.7112	5.14 = AB

(d) In the afternoon, when the angle of elevation of the sun is 25°, the shadow of a tree is 60 feet longer than it is at noon, when the sun's angle of elevation is 50°. How high is the tree?

Let AB represent the tree, AN the afternoon shadow, and AM the shadow at noon. Then, from the right triangle MAB, we have

(7) $$AM = AB \cot 50°$$

and, from the right triangle NAB,

(8) $$AN = AB \cot 25°.$$

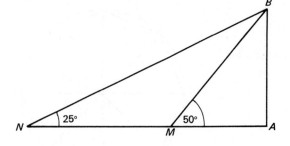

Figure 5.3

Subtract (7) from (8):

$$AN - AM = AB(\cot 25° - \cot 50°).$$

But

$$AN - AM = MN$$
$$= 60 \text{ feet,}$$

and therefore the height of the tree is

$$AB = \frac{60}{\cot 25° - \cot 50°}$$

$$= \frac{60}{\tan 65° - \tan 40°}$$

$$= \frac{60}{2.145 - .8391}$$

$$= \frac{60}{1.306}$$

$$= 45.9 \text{ feet.}$$

In map reading, navigation, etc., the *direction* of a line *AB* (or the *bearing* of *AB*) is defined to be the smaller of the two angles $\angle BAN$, $\angle BAS$, where *AN* is due north and *AS* is due south.

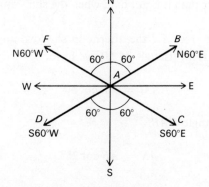

Figure 5.4

For example, in Figure 5.4 the direction of AB is 60° east of north, that of AC is 60° east of south, that of AD is 60° west of south, and that of AF is 60° west of north. These directions are denoted by N60°E, S60°E, S60°W, and N60°W, respectively.

Example 1.2 (a) Newport is 123 miles due west of London and 141 miles due south of Southport. What is the distance and direction from London to Southport? [All distances are measured as the crow flies; the effect of curvature of the Earth is ignored.]

Figure 5.5

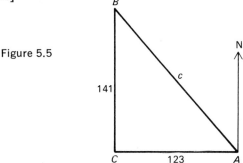

In Figure 5.5, the points A, B, C represent London, Southport, and Newport, respectively. The problem is to determine $\angle NAB$, that is, angle B, and the distance corresponding to c. In the right-angled triangle ABC,

$$\tan B = \frac{b}{a}$$

$$= \frac{123}{141}.$$

Numbers	Logs	Antilogs
123	2.0899	
÷ 141	− 2.1492	
	$\overline{1}.9407$	
	$= \log \tan B$	$41° = B$

Hence Southport is N41°W of London. The required distance can be obtained by an application of the Pythagorean theorem. However, we shall compute it by means of formula (2):

$$c = \frac{a}{\cos B}$$

$$= \frac{141}{\cos 41°}.$$

Numbers	Logs	Antilogs
141	2.1492	
÷ cos 41°	− $\overline{1}$.8778	
	2.2714	187 = c

The distance from London to Southport is 187 miles.

(b) Jerusalem and Haifa are respectively 54 km S53°E and 86 km N13°E of Tel Aviv. A plane flies from Jerusalem to Haifa in a straight line. What is the direction of the flight? How close will the plane come to Tel Aviv? What is the distance flown?

(We shall obtain in Section 5.2 a formula which will allow us to solve a problem of this type more efficiently. Here we use elementary methods of solution.)

Figure 5.6

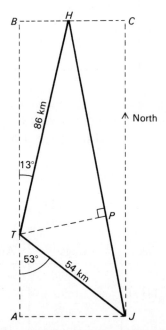

In Figure 5.6 points *J*, *H*, *T* represent Jerusalem, Haifa, and Tel Aviv, respectively. From *T* draw the perpendicular to *JH* intersecting *JH* at *P*. *JC* and *AB* are drawn in the South-North direction, *AJ* and *BC* in the West-East direction. In the right-angled triangle *TAJ*,

$$AJ = TJ \sin (\angle ATJ)$$
$$= 54 \sin 53°,$$
$$AT = TJ \cos (\angle ATJ)$$
$$= 54 \cos 53°.$$

Similarly, from the right-angled triangle TBH,

$$BH = TH \sin (\angle BTH)$$
$$= 86 \sin 13°,$$
$$TB = TH \cos (\angle BTH)$$
$$= 86 \cos 13°.$$

Numbers	Logs	Antilogs
54	1.7324	
$\times \sin 53°$	$+ \bar{1}.9024$	
	1.6348	$43.1 = AJ$
54	1.7324	
$\times \cos 53°$	$+ \bar{1}.7795$	
	1.5119	$32.5 = AT$
86	1.9345	
$\times \sin 13°$	$+ \bar{1}.3521$	
	1.2866	$19.3 = BH$
86	1.9345	
$\times \cos 13°$	$+ \bar{1}.9887$	
	1.9232	$83.8 = TB$

Thus

$$JC = AB = AT + TB$$
$$= 116 \text{ km, approximately};$$
$$HC = BC - BH = AJ - BH$$
$$= 24 \text{ km, approximately}.$$

Hence from the right-angled triangle JCH:

$$\tan (\angle HJC) = \frac{HC}{JC}$$
$$= \frac{24}{116} .$$

Numbers	Logs	Antilogs
24	1.3802	
$\div$ 116	$-$ 2.0645	
	$\bar{1}.3157$	
	$= \log \tan (\angle HJC)$	$11°30' = \angle HJC$

Thus the direction of flight is N11°30′W. The shortest distance that the plane passes from Tel Aviv is (from right-angled triangle *JPT*)

$$PT = JT \sin (\angle PJT)$$
$$= JT \sin \left(90° - \angle HJC - (90° - \angle ATJ)\right)$$
$$= 54 \sin 41°30′;$$

and the distance from Jerusalem to Haifa is (from right-angled triangle *JCH*)

$$JH = \frac{HC}{\sin (\angle HJC)}$$

$$= \frac{24}{\sin 11°30′} \cdot$$

Numbers	Logs	Antilogs
54	1.7324	
× sin 41°30′	+ $\overline{1}$.8213	
	1.5537	36 = *PT*
24	1.3802	
÷ sin 11°30′	− $\overline{1}$.2997	
	2.0805	120 = *JH*

SYNOPSIS

In this section we reviewed methods of solving right-angled triangles and used logarithms to facilitate the solution of various examples, including some involving geographic directions.

QUIZ

Answer *true* or *false* (Questions 1–8 refer to any triangle *ABC* right-angled at *C*):
1. $\sin A = \cos B$.
2. $a \sin A = b \sin B$.
3. $a \cos A = b \cos B$.
4. $a \sin A + b \sin B = c$.
5. $\cos (A + B) = \cos C$.
6. $(c^2 - a^2) + (c^2 - b^2) = a^2 + b^2$.
7. If $A \neq B$, then $c^2 - a^2 \neq a^2 - b^2$.
8. If $A = B$, then $c^2 - a^2 \neq a^2 - b^2$.

9. If X and Y are N45°E and S45°E of Z, respectively, and $XZ = YZ$ then X is due north of Y.

10. If P is N35°W of Q and R is S35°E of P, then R is N35°W of Q.

EXERCISES

1. Let ABC be a triangle right-angled at C. Let h be the length of the altitude CD drawn from the vertex C, and let S be the area of ABC. [Recall that $S = \frac{1}{2}ab = \frac{1}{2}ch$.]

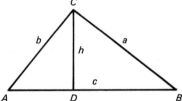

Figure 5.7

Solve triangle ABC, given that:

(a) $a = 7$, $A = 32°$;

(b) $c = 14.2$, $A = 47°$;

(c) $b = 27.6$, $B = 49°$;

(d) $a = 3.92$, $c = 5.4$;

(e) $c = 65.6$, $b = 42.3$;

(f) $c = 243$, $B = 25°$;

(g) $a = 43$, $b = 28$;

(h) $a = 3.05$, $B = 31°30'$;

(i) $a = 3.26$, $B = 66°30'$;

(j) $b = 5$, $h = 4$;

(k) $a = 13$, $h = 12$;

(l) $a = 6.72$, $h = 4.18$;

(m) $h = 4.7$, $A = 39°$;

(n) $S = 65$, $a = 10$;

(o) $S = 11.4$, $b = 4.6$;

(p) $S = 24.3$, $a = 7.9$;

(q) $S = 50$, $A = 35°$

[*Hint:* $S = \frac{1}{2}ab$ and $a = b \tan A$ imply $S = \frac{1}{2}b^2 \tan A$. Thus

$$b = \sqrt{\frac{2S}{\tan A}} \quad \text{and} \quad a = \sqrt{2S \tan A}\,]\,;$$

(r) $S = 6.35$, $B = 49°$;

(s) $S = 14.2$, $A = 31°30'$.

2. Find the perimeter of a regular n-sided polygon inscribed in a circle of radius r, if:

(a) $n = 10$, $r = 5.35$;

(b) $n = 5$, $r = 22.5$;

(c) $n = 12$, $r = 9.64$;

(d) $n = 20$, $r = 0.75$.

3. Find the radius of the circle circumscribed on an n-sided regular polygon whose perimeter is p, if:

(a) $n = 8$, $p = 26.8$;

(b) $n = 15$, $p = 250$;

(c) $n = 10$, $p = 1.25$;

(d) $n = 18$, $p = 27$.

4. The lengths of the diagonals of a rhombus are 3.68 and 4.82 inches. What are the angles and the length of a side of the rhombus?
5. What is the angle between the diagonal of a cube and a diagonal of a face of the cube that meets it?
6. A plane P has an angle of elevation of 33° from a point X on a runway and an elevation of 40°30′ from a point Y on the runway. If $XY = 576$ ft., find the height of the plane, i.e., the distance from P to the point Z directly below it, assuming that X, Y, Z lie in a horizontal plane and $\angle XZP = 90°$. [*Hint:* Let h be the required height; using h evaluate XZ and YZ from triangles XZP and YZP; use the Pythagorean theorem.]
7. A driver travelling due south observes two stationary vehicles on a road due east. After a minute's driving, the directions of the vehicles are observed to be N17°E and N27°E. If the distance between the stationary vehicles is 300 yds., what is the speed of the car?
8. Lisbon and Madrid are respectively N59°W and N31°E of Sevilla. If Madrid is 314 mi. N69°30′E of Lisbon, what are the distances between Lisbon and Sevilla, and between Sevilla and Madrid?
9. Malaga is 99 mi. S59°W of Sevilla. Using the data of Exercise 8, find the direction and the distance of Malaga from Madrid.

5.2   Oblique Triangles

In this section we tackle the problem of solving oblique triangles, that is, triangles which are not necessarily right-angled. The three angles and the three sides of a triangle are called the *parts* of the triangle. A triangle can be solved if enough parts are given so that the triangle can be constructed, that is, at least three out of the six parts, not all three being angles. Recall that there are three standard cases: A triangle is determined (i.e., can be constructed unambiguously) if either

(i)   two angles and a side are given, or
(ii) two sides and the included angle are given, or
(iii) all three sides are given.

We shall also discuss the so-called *ambiguous case* when two sides and the angle opposite one of them are given.

We can solve an oblique triangle by subdividing it into right-angled triangles or, as we did in Example 1.2 (b), by constructing other auxiliary right-angled triangles. Instead of performing this chore ad hoc for each triangle, we derive several general formulas that allow us to routinely solve any adequately defined oblique triangle.

Let ABC be any triangle. As before, we denote the angles $\angle BAC$, $\angle ABC$, $\angle ACB$ by A, B, C, respectively. Denote the lengths BC, CA, AB by a, b, c, respectively.

Theorem 2.1 (Law of Sines) *If the sides and angles in a triangle ABC are denoted as above, then*

$$\frac{a}{\sin A} = \frac{b}{\sin B} = \frac{c}{\sin C}.$$

Proof First suppose that B and C are both acute angles. Let AD be the altitude and let h be the length of AD.

Figure 5.8

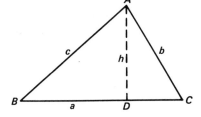

In triangle ABD,

$$h = c \sin B,$$

and in triangle ADC,

$$h = b \sin C.$$

Thus,

$$b \sin C = c \sin B,$$

or, since $\sin C$ and $\sin B$ are nonzero,

$$\frac{b}{\sin B} = \frac{c}{\sin C}.$$

Next suppose that either B or C is obtuse, say C. Draw the altitude AD and let h be the length of AD.

Figure 5.9

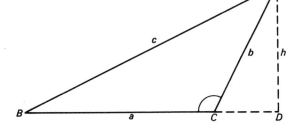

Then from triangle ACD,

$$h = b \sin (\angle ACD).$$

But $\angle ACD = 180° - C$, and therefore

$$h = b \sin (180° - C)$$
$$= b \sin C.$$

Also, from triangle ABD,

$$h = c \sin B.$$

Hence, we again have

$$\frac{b}{\sin B} = \frac{c}{\sin C}.$$

Lastly, if C is a right angle, then

$$\sin B = b/c,$$

and thus, since $\sin C = 1$,

$$\frac{b}{\sin B} = \frac{c}{\sin C}.$$

We have proved that for any triangle ABC,

$$\frac{b}{\sin B} = \frac{c}{\sin C}.$$

Similarly we can show that

$$\frac{a}{\sin A} = \frac{b}{\sin B},$$

and the result follows. ▌

The law of sines is an important formula for solving triangles, particularly when two angles and a side are given.

Example 2.1 (a) Solve the triangle ABC in which $A = 47°$, $B = 72°$, $a = 4$. Using the law of sines, we calculate that

$$b = \frac{a}{\sin A} \cdot \sin B$$

$$= \frac{4}{\sin 47°} \cdot \sin 72°.$$

Numbers	Logs	Antilogs
4	0.6021	
$\times \sin 72°$	$+ \bar{1}.9782$	
	0.5803	
$\div \sin 47°$	$- \bar{1}.8641$	
	0.7162	$5.20 = b$

Hence,

$$b = 5.20.$$

Since the sum of the three angles of a triangle is 180°, we have

$$C = 180° - (A + B)$$
$$= 61°;$$

therefore,

$$c = \frac{a \sin C}{\sin A}$$
$$= \frac{4 \sin 61°}{\sin 47°}.$$

Numbers	Logs	Antilogs
4	0.6021	
$\times \sin 61°$	$+ \bar{1}.9418$	
	0.5439	
$\div \sin 47°$	$- \bar{1}.8641$	
	0.6798	$4.78 = c$

Hence,

$$c = 4.78.$$

(b) Paris is 259 miles S24°W of Amsterdam. If the direction from Paris to London is N30°W and the direction from London to Amsterdam is N75°E, find the distances from London to Amsterdam and Paris.

The three cities are represented in Figure 5.10 by their initial letters; the lines with arrows represent the South-North direction. By the law of sines,

$$\frac{LA}{\sin (\angle APL)} = \frac{AP}{\sin (\angle PLA)} = \frac{PL}{\sin (\angle LAP)}.$$

Figure 5.10

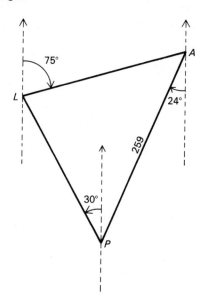

It is easy to see that $\angle APL = 54°$, $\angle PLA = 75°$ and $\angle LAP = 51°$. Since $AP = 259$ mi., we have

$$LA = \frac{259}{\sin 75°} \sin 54°,$$

$$PL = \frac{259}{\sin 75°} \sin 51°.$$

Numbers	Logs	Antilogs
259	2.4133	
÷ sin 75°	− $\overline{1}$.9849	
	2.4284	
× sin 54°	+ $\overline{1}$.9080	
	2.3364	217 = LA
259	2.4133	
÷ sin 75°	− $\overline{1}$.9849	
	2.4284	
× sin 51°	+ $\overline{1}$.8905	
	2.3189	208 = PL

Hence it is 217 miles from London to Amsterdam, and 208 mi. from London to Paris.

The law of sines is also used in the "ambiguous case" in which two sides and the angle opposite one of them are given. It is well known in geometry that in this case there may be two distinct solutions (i.e., there may be two distinct triangles that satisfy the given conditions), one solution or no solution. These three situations are illustrated in Figures 5.11 (i), (ii), and (iii), respectively; in all cases a, b, and A are given.

Figure 5.11

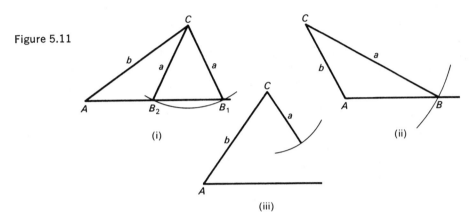

(i)

(iii)

(ii)

We further illustrate the ambiguous case in the following examples.

Example 2.2 (a) Solve the triangle ABC in which $A = 46°$, $a = 4$, $b = 5$.
We compute that

$$\sin B = \frac{b \sin A}{a}$$

(1)
$$= \frac{5 \sin 46°}{4}$$

$$= \tfrac{5}{4}(0.7193)$$

$$= 0.8991.$$

We can then find from the tables that

$$B = 64°.$$

In the interval $(0°, 180°)$, equation (1) has another solution:

$$B = 180° - 64°$$
$$= 116°.$$

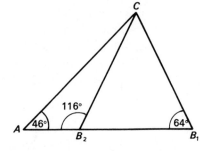

Figure 5.12

Thus there are two possible solutions. (See Figure 5.12.) We continue to solve the triangle separately for the alternative cases, $B_1 = 64°$ and $B_2 = 116°$.

Case 1: If $B_1 = 64°$, then $C_1 = 180° - (46° + 64°) = 70°$. Thus the length of the side AB_1 is

$$c_1 = a \frac{\sin C_1}{\sin A}$$

$$= 4 \frac{\sin 70°}{\sin 46°} .$$

Numbers	Logs	Antilogs
4	0.6021	
$\times \sin 70°$	$+ \bar{1}.9730$	
	0.5751	
$\div \sin 46°$	$- \bar{1}.8569$	
	0.7182	$5.23 = c_1$

Thus

$$c_1 = 5.23.$$

Case 2: If $B_2 = 116°$, then $C_2 = 180° - (116° + 46°) = 18°$, and

$$c_2 = a \frac{\sin C_2}{\sin A}$$

$$= 4 \frac{\sin 18°}{\sin 46°} .$$

Numbers	Logs	Antilogs
4	0.6021	
$\times \sin 18°$	$+ \bar{1}.4900$	
	0.0921	
$\div \sin 46°$	$- \bar{1}.8569$	
	0.2352	$1.72 = c_2$

Hence,
$$c_2 = 1.72.$$

(b) Solve the triangle ABC in which $A = 51°$, $a = 4$, $b = 3$.
By the law of sines,
$$\frac{a}{\sin A} = \frac{b}{\sin B}.$$

Therefore

$$\sin B = \frac{b \sin A}{a}$$
$$= \frac{3 \sin 51°}{4}$$
$$= \tfrac{3}{4}(0.7771)$$
(2) $$= 0.5828.$$

We find from the tables that with accuracy to $30'$,

$$\sin 35°30' = 0.5828.$$

We know from Theorem 2.1, Section 3.2, that

$$\sin (180° - 35°30') = \sin 35°30'.$$

Thus (2) has two solutions in the interval $(0°, 180°)$, namely, $B = 35°30'$ or $144°30'$. However, the angles in a triangle add up to $180°$ and we know that $A = 51°$. Hence angle B cannot be $144°30'$. We therefore have

$$B = 35°30'.$$

It follows that
$$C = 180° - (A + B)$$
$$= 180° - (51° + 35°30')$$
$$= 93°30'.$$

We complete our solution by computing

$$c = a \frac{\sin C}{\sin A}$$
$$= 4 \frac{\sin 93°30'}{\sin 51°}.$$

Numbers	Logs	Antilogs
4	0.6021	
$\times \sin 93°30'$		
$(= \sin 86°30')$	$+ \overline{1}.9992$	
	0.6013	
$\div \sin 51°$	$- \overline{1}.8905$	
	0.7108	$5.14 = c$

Therefore

$$c = 5.14.$$

(c)   Solve the triangle ABC in which $a = 4$, $b = 6$, $A = 62°$. If there is a solution, then

$$\frac{a}{\sin A} = \frac{b}{\sin B},$$

i.e.,

$$\sin B = \frac{b}{a} \sin A$$
$$= \tfrac{6}{4} \sin 62°$$
$$= \tfrac{3}{2} 0.8830$$
$$= 1.3245.$$

But this is impossible since $-1 \leq \sin \theta \leq 1$ for any angle θ.

We now derive important formulas that generalize the Pythagorean theorem. We shall use these formulas to solve triangles in the case when two sides and the included angle or all three sides are given.

Theorem 2.2   (Law of Cosines)   *If ABC is a triangle, then, in the notation of the preceding theorem,*

(3) $$c^2 = a^2 + b^2 - 2ab \cos C,$$

(4) $$a^2 = b^2 + c^2 - 2bc \cos A,$$

(5) $$b^2 = c^2 + a^2 - 2ca \cos B.$$

Proof   We prove formula (3).   Orient the triangle with respect to a set of coordinate axes so that the vertices C, A, B are the points $(0, 0)$, $(b, 0)$, $(a \cos C, a \sin C)$, respectively.   For the case where C is acute see Figure 5.13, and for the case where C is obtuse see Figure 5.14.

Figure 5.13

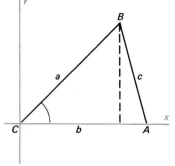

Figure 5.14

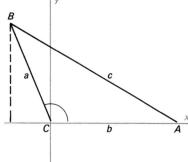

We now use the distance formula (see Theorem 3.1, Chapter 2), to compute the distance from A to B:

$$\begin{aligned}
c^2 &= (b - a \cos C)^2 + (-a \sin C)^2 \\
&= b^2 - 2ab \cos C + a^2 \cos^2 C + a^2 \sin^2 C \\
&= a^2(\sin^2 C + \cos^2 C) + b^2 - 2ab \cos C \\
&= a^2 + b^2 - 2ab \cos C.
\end{aligned}$$

Note that the distance formula $\big($and thus (3)$\big)$ is valid regardless of the quadrants in which the points happen to be located. Formulas (4) and (5) are proved similarly. ∎

Example 2.3 (a) In the triangle ABC, $C = 110°$, $a = 10$, and $b = 5$. Find the remaining parts of the triangle.

Using the law of cosines, we have

$$\begin{aligned}
c^2 &= 10^2 + 5^2 - 2 \cdot 10 \cdot 5 \cdot \cos 110° \\
&= 10^2 + 5^2 + 2 \cdot 10 \cdot 5 \cdot \cos 70° \\
&= 125 + 100 \times 0.3420 \\
&= 159.2.
\end{aligned}$$

Hence,

$$c = 12.6.$$

Now, by the law of sines,

$$\frac{a}{\sin A} = \frac{c}{\sin C},$$

i.e.,

$$\sin A = \frac{10 \sin 110°}{12.6}.$$

Numbers	Logs	Antilogs
10	1.0000	
$\times \sin 110°$	$+ \overline{1}.9730$	
	0.9730	
$\div 12.6$	$- 1.1004$	
	$\overline{1}.8726$	$\sin 48°$

Note that although $\sin 132° = \sin 48°$, the angle A cannot be obtuse, since the triangle already has an obtuse angle C. Hence, $A = 48°$. Lastly, we compute

$$B = 180° - (A + C)$$
$$= 22°.$$

(b)   In triangle ABC, $a = 5$, $b = 7$, and $c = 4$. Compute B. By the law of cosines

$$b^2 = a^2 + c^2 - 2ac \cos B,$$

and therefore

$$\cos B = \frac{a^2 + c^2 - b^2}{2ac}$$
$$= \frac{5^2 + 4^2 - 7^2}{2 \times 5 \times 4}$$
$$= -8/40$$
$$= -0.2000.$$

We find from the tables that $\cos 78°30' = 0.2000$ (approximately). Since $\cos B = -0.2000$, we can conclude that

$$B = 180° - 78°30'$$
$$= 101°30'.$$

(c)   A United Nations Organization mediator wants to go from Tel Aviv to Cairo, a distance of 250 miles in the direction S57°W. Since there is no direct airline connection between the two cities, the mediator has to fly from Tel Aviv to Athens (direction N55°W) and then from Athens to Cairo (direction S35°30'E). What is the total distance the mediator covers flying from Tel Aviv to Cairo via Athens?

We ignore the effect of the earth's curvature. The problem then is to compute two sides of the Tel Aviv–Athens–Cairo triangle (triangle TAC). We sketch the triangle and find, using elementary geometry, that $\angle ATC = 68°$, $\angle CAT = 19°30'$, and $\angle TCA = 92°30'$.

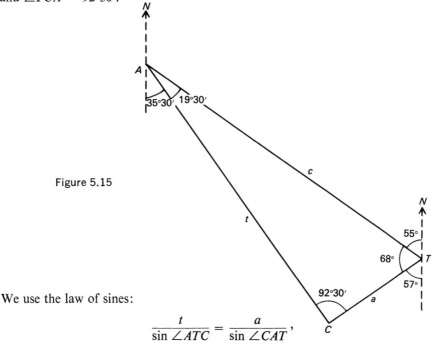

Figure 5.15

We use the law of sines:

$$\frac{t}{\sin \angle ATC} = \frac{a}{\sin \angle CAT},$$

i.e.,

$$t = 250 \, \frac{\sin 68°}{\sin 19°30'}.$$

Numbers	Logs	Antilogs
250	2.3979	
× sin 68°	+ $\overline{1}$.9672	
	2.3651	
÷ sin 19°30'	− $\overline{1}$.5235	
	2.8416	694 = t

Therefore $t = 694$ miles.

Now, for a change, we use the law of cosines to compute the distance from Tel Aviv to Athens:

$$c^2 = a^2 + t^2 - 2at \cos C$$
$$= 250^2 + 694^2 - 2 \cdot 250 \cdot 694 \cdot \cos 92°30'$$
$$= 559,000.$$

(We have omitted the details of computation.) Thus, approximating,

$$c = \text{antilog} \left(\tfrac{1}{2} \log 559000 \right)$$
$$= \text{antilog } 2.8737$$
$$= 748.$$

Hence the required distance is

$$c + t = 694 + 748$$
$$= 1,442 \text{ miles.}$$

We conclude this section with a theorem that provides an alternative method, more suitable for logarithmic computations than the law of cosines, for solving a triangle, given two of its sides and the included angle.

Theorem 2.3   (Law of Tangents)   *In any triangle ABC,*

(6)
$$\frac{a - b}{a + b} = \frac{\tan \dfrac{A - B}{2}}{\tan \dfrac{A + B}{2}}.$$

(Analogous formulas hold for the other two pairs of sides.)

Proof   By Theorem 2.1,

$$\frac{a}{b} = \frac{\sin A}{\sin B};$$

therefore

$$\frac{a}{b} + 1 = \frac{\sin A}{\sin B} + 1,$$

i.e.,

(7)
$$\frac{a + b}{b} = \frac{\sin A + \sin B}{\sin B}.$$

Similarly,

(8)
$$\frac{a - b}{b} = \frac{\sin A - \sin B}{\sin B}.$$

Divide (8) by (7) to obtain

(9)
$$\frac{a - b}{a + b} = \frac{\sin A - \sin B}{\sin A + \sin B}.$$

Now, by formulas (28) and (27) of Section 4.2,

(10)
$$\frac{\sin A - \sin B}{\sin A + \sin B} = \frac{2 \sin \dfrac{A - B}{2} \cos \dfrac{A + B}{2}}{2 \cos \dfrac{A - B}{2} \sin \dfrac{A + B}{2}}$$

$$= \frac{\tan \dfrac{A - B}{2}}{\tan \dfrac{A + B}{2}}.$$

Combining (9) and (10) we obtain (6). ∎

Note that $A + B + C = 180°$ and therefore

(11)
$$\frac{A + B}{2} = \frac{180° - C}{2}$$

$$= 90° - \frac{C}{2}.$$

Thus if a, b, and C are given, we can use formula (6) to compute $\dfrac{A - B}{2}$, which together with (11) allows us to evaluate A and B.

Example 2.4 In a triangle ABC, given $a = 5.72$, $b = 3.31$, $C = 100°$, find the remaining parts of the triangle.

We substitute in (6):

$$\frac{5.72 - 3.31}{5.72 + 3.31} = \frac{\tan \dfrac{A - B}{2}}{\tan \left(90° - \dfrac{100°}{2} \right)}.$$

Thus

$$\tan \frac{A - B}{2} = \frac{2.41}{9.03} \tan 40°.$$

Numbers	Logs	Antilogs
2.41	0.3820	
× tan 40°	+ $\overline{1}$.9238	
	0.3058	
÷ 9.03	− 0.9557	
	$\overline{1}$.3501	12°30′ = $\dfrac{A - B}{2}$

Therefore, since $A + B = 180° − C = 80°$, $\frac{1}{2}A + \frac{1}{2}B = 40°$, we have

$$\frac{1}{2}A - \frac{1}{2}B = 12°30′,$$
$$\frac{1}{2}A + \frac{1}{2}B = 40°.$$

Solving these two equations we obtain

$$A = 52°30′,$$
$$B = 27°30′.$$

We use the law of sines to compute the third side:

$$c = a \frac{\sin C}{\sin A}$$

$$= 5.72 \frac{\sin 100°}{\sin 52°30′}.$$

Numbers	Logs	Antilogs
5.72	0.7574	
× sin 100°		
(= sin 80°)	+ $\overline{1}$.9934	
	0.7508	
÷ sin 52°30′	− $\overline{1}$.8995	
	0.8513	7.10 = c

SYNOPSIS

In this section we derived three important formulas that hold for any triangle *ABC*:

$$\frac{a}{\sin A} = \frac{b}{\sin B} \qquad \text{(Law of Sines)}$$

$$c^2 = a^2 + b^2 - 2ab \cos C \quad \text{(Law of Cosines)}$$

$$\frac{a - b}{a + b} = \frac{\tan \dfrac{A - B}{2}}{\tan \dfrac{A + B}{2}} \qquad \text{(Law of Tangents)}$$

and showed how to apply them to solve any triangle.

QUIZ

Answer *true* or *false* (Questions 2–10 refer to any triangle ABC):

1. If Z is due east of X, and if Y is N60°E of X and N60°W of Z, then $\angle XYZ = 60°$.

2. $\dfrac{\sin A}{a} = \dfrac{\sin B}{b}$.

3. If $A = 2B$, then $a = 2b$.

4. $\tan \dfrac{A + B}{2} = \cot \dfrac{C}{2}$.

5. If $\sin A = \sin B$, then $A = B$.

6. $\dfrac{a}{\sin (B + C)} = \dfrac{b}{\sin (A + C)}$.

7. $a^2 = b^2 + c^2 - 2bc \cos (B + C)$.

8. The angle C is acute if and only if $c < a + b$.

9. The angle C is obtuse if and only if $c^2 < a^2 + b^2$.

10. The angle C is acute if and only if $c^2 < a^2 + b^2$.

EXERCISES

1. In each of the following, solve triangle ABC:
   (a) $A = 48°$, $B = 74°$, $c = 10$;
   (b) $A = 38°$, $B = 44°$, $a = 1$;
   (c) $B = 37°$, $C = 100°$, $a = 17.6$;
   (d) $B = 97°$, $C = 62°$, $b = 2.63$;
   (e) $A = 31°30'$, $C = 90°30'$, $b = 0.034$;
   (f) $A = 42°30'$, $C = 90°$, $b = 0.672$;
   (g) $a = 5$, $b = 7$, $c = 10$;
   (h) $a = 12.5$, $b = 7.5$, $c = 10$;

  (i) $a = 5.06, b = 2.61, c = 3.12$;

  (j) $a = 156, b = 173, c = 173$;

  (k) $a = 39.6, b = 42.9, c = 16.5$;

  (l) $a = 2, b = 3, C = 40°$;

  (m) $b = 17, c = 39, A = 73°30'$;

  (n) $c = 25.9, a = 16.7, B = 23°$;

  (o) $b = 3.46, c = 5.12, A = 108°30'$;

  (p) $a = 0.183, b = 0.217, C = 90°$;

  (q) $a = 2.35, b = 2.35, C = 129°$;

  (r) $A = 13°30', b = 653, c = 327$.

2. In each of the following find all possible solutions (if any):

  (a) $a = 5, b = 4, B = 50°$;

  (b) $a = 5, b = 4, B = 130°$;

  (c) $a = 5, b = 7, B = 70°$;

  (d) $A = 103°, a = 78.3, c = 61.3$;

  (e) $A = 44°, a = 1.93, b = 2.51$;

  (f) $C = 20°, b = 150, c = 51.3$;

  (g) $A = 110°, b = 97.2, a = 94.9$;

  (h) $a = 3.27, b = 8.67, c = 4.91$.

3. The diagonals of a parallelogram are 13 and 19 units of length. They intersect at 48°. Find the angles and the lengths of sides of the parallelogram.

4. The lengths of adjacent sides of a parallelogram are 13 and 7 units, while one of its angles is 110°. Find the lengths of the diagonals of the parallelogram.

5. The sides of a triangle are 17, 21, and 23 units of length. What are the lengths of the altitudes?

6. A jet plane flies over town X and heads in a direction N20°W at 600 m.p.h. At the same time, a propeller plane passes over town Y and heads in a direction N50°E at a speed of 270 m.p.h. The two planes meet after 40 minutes; find the distance from X to Y.

7. Show that in any triangle ABC, $a^2 + b^2 + c^2 = 2bc \cos A + 2ca \cos B + 2ab \cos C$.

8. Prove that the area of triangle ABC is $\frac{1}{2}ab \sin C$.

  [*Hint:* Assume that the area of a triangle is equal to one half the product of a side and the corresponding altitude.]

9. In parallelogram $PQRS$, $\angle PQR = \theta$, $PQ = x$, and $QR = y$. Show that the area of $PQRS$ is $xy \sin \theta$.

10. Show that the area of triangle ABC is

$$\frac{1}{2} c^2 \frac{\sin A \sin B}{\sin C}.$$

[*Hint:* Recall that the area S of ABC is given by $S = \frac{1}{2}ch$ (see Figure 5.16).

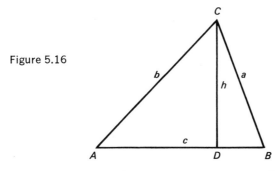

Figure 5.16

By the law of sines, $\dfrac{b}{\sin B} = \dfrac{c}{\sin C}$, i.e., $b = c \dfrac{\sin B}{\sin C}$. Also, from tri-

angle ADC, $h = b \sin A$. Thus $h = c \dfrac{\sin A \sin B}{\sin C}$.]

11. Compute the area of triangle ABC if:
    (a) $b = 4$, $c = 5$, $A = 49°$;
    (b) $b = 6$, $A = 50°$, $B = 70°$;
    (c) $c = 10$, $A = 75°$, $B = 100°$;
    (d) $a = 3.93$, $A = 32°$, $B = 78°30'$;
    (e) $b = 137$, $A = 109°$, $C = 42°$;
    (f) $b = 12.9$, $A = 24°$, $C = 34°30'$.

12. A plane flies in a straight line from Marseille to Genova and from Genova to Ancona. The pilot wants to save time and expense on the return journey by flying directly in a straight line from Ancona to Marseille. If Genova is 190 mi. N63°E of Marseille and Ancona is 230 mi. S75°E of Genova, in what direction should the pilot fly on the return journey? If the speed of his plane (assumed constant) is 200 m.p.h., how much flight time will the pilot save?

13. Istanbul is 307 mi. N85°E of Salonica. Find the directions from Istanbul to Athens and from Athens to Salonica, given that distances from Athens to Istanbul and to Salonica are 343 mi. and 185 mi., respectively.

5.3 Vectors*

A vector is one of the most ubiquitous objects in mathematics and physics. We shall not attempt to develop a general theory of vectors here. We shall only demonstrate how trigonometry can be applied to solve certain types of problems involving vector quantities that are encountered in elementary physics.

*It is assumed that the student is acquainted with the definitions and elementary properties of forces and velocities. If this is not the case, the section may be omitted without loss of continuity.

If a quantity is completely determined by its magnitude (i.e., the number of appropriate units of measurement), it is called a *scalar* magnitude. For example, length, area, temperature, mass, and energy are scalar quantities. Other quantities have both magnitude and direction and are not determined until the two are specified. These are called *directed* quantities. Examples of directed quantities are displacement, velocity, acceleration, force, and rotation. [*Note:* The student is cautioned that directed quantities in many textbooks are incorrectly called vectors; there are directed quantities which are definitely not vectors.]

A directed quantity can often be conveniently represented by a directed line segment (or an arrow) whose length and direction represent the magnitude and the direction of the directed quantity. We shall denote both a directed quantity and its geometric representation by the same letter topped with an arrow, e.g., $\vec{v}$; or by two capital letters topped with an arrow, e.g., $\overrightarrow{AB}$, where A is the *initial point* and B the *terminal point* of the quantity. The magnitude of $\vec{v}$ is denoted by v and that of $\overrightarrow{AB}$ by AB.

Example 3.1   Represent geometrically forces of 3 lbs., 2 lbs., and 4 lbs. acting in the directions N30°E, S40°E, and N55°W, respectively.

We adopt the usual cartographic convention as to geographic directions and choose a scale of 1 in. to represent 2 lbs. of force. We use a common initial point for the three quantities. In Figure 5.17, $\overrightarrow{OA}$, $\overrightarrow{OB}$, and $\overrightarrow{OC}$ respectively represent the three forces.

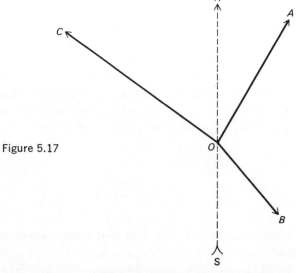

Figure 5.17

Definition 3.1   (Vectors)   Directed quantities of the same kind, together with operations of *scalar multiplication* and *vector addition* as defined below, are called (real) *vectors.*

(i) If a is a positive number and $\vec{v}$ a vector, then the *scalar* product $a\vec{v}$ is the vector whose direction is that of $\vec{v}$ and whose magnitude is av. If $a < 0$, then $a\vec{v}$ is the vector whose direction is opposite to that of $\vec{v}$ (i.e., differing from the direction of $\vec{v}$ by 180°), and whose magnitude is $-av$. If $a = 0$, then $a\vec{v} = \vec{0}$, the *zero vector*, a convenient conventional object of zero magnitude and no direction.

(ii) If $\overrightarrow{OA}$ and $\overrightarrow{OB}$ are noncollinear vectors, then their *(vector) sum*, or their *resultant*, is the vector $\overrightarrow{OC}$ where $OACB$ is a parallelogram. If $\overrightarrow{OA}$ and $\overrightarrow{OB}$ are collinear then their resultant $\overrightarrow{OC}$ is collinear with both vectors and its magnitude is $OA + OB$ or $OA - OB$, according as $\overrightarrow{OA}$ and $\overrightarrow{OB}$ are in the same or in opposite directions. In all cases, if $\overrightarrow{OC}$ is the resultant of $\overrightarrow{OA}$ and $\overrightarrow{OB}$, we write $\overrightarrow{OC} = \overrightarrow{OA} + \overrightarrow{OB}$.

Figure 5.18
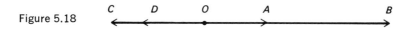

These definitions are illustrated in Figures 5.18 and 5.19. In Figure 5.18, $\overrightarrow{OB} = 3\overrightarrow{OA}$, $\overrightarrow{OC} = -\frac{3}{2}\overrightarrow{OA}$, $\overrightarrow{OD} = -1\overrightarrow{OA}$. It is customary to write $-\overrightarrow{OA}$ instead of $-1\overrightarrow{OA}$. Thus $-\overrightarrow{OA}$ is collinear with $\overrightarrow{OA}$, of equal magnitude and opposite direction.

Figure 5.19
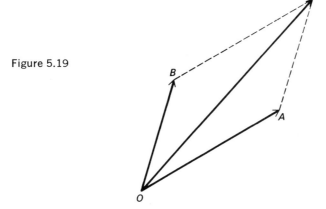

Example 3.2 The angle between vectors $\overrightarrow{OA}$ and $\overrightarrow{OB}$ is 43° and their magnitudes are 10 and 7 units, respectively. Find the resultant $\overrightarrow{OC} = \overrightarrow{OA} + \overrightarrow{OB}$, that is, compute the magnitude OC and $\angle AOC$, the angle that the resultant makes with one of the given vectors.

The two vectors are represented in Figure 5.19 where 1 in. represents 4 units. Since $OACB$ is a parallelogram, $AC = OB = 7$ and $\angle OAC = 180° - \angle AOB = 180° - 43° = 137°$. Our problem is therefore to solve triangle OAC, given two sides and the included angle. We first use the law of cosines:

$$OC^2 = OA^2 + AC^2 - 2\,OA \cdot AC \cos{(\angle OAC)}$$
$$= 10^2 + 7^2 - 2 \times 10 \times 7 \cos 137°$$
$$= 100 + 49 + 140 \cos 43°$$
$$= 149 + 140 \times 0.7314$$
$$= 251,$$

to three significant figures. We compute

Numbers	Logs	Antilogs
$\sqrt{251}$	$\frac{1}{2} \times 2.3997$ $= 1.1998$	$15.8 = OC$

Now apply the law of sines:

$$\sin{(\angle AOC)} = \frac{AC}{OC} \sin{(\angle OAC)}$$

$$= \frac{7}{15.8} \sin 137°.$$

Numbers	Logs	Antilogs
7	0.8451	
$\times \sin 137°$		
$(= \sin 43°)$	$+ \overline{1}.8338$	
	0.6789	
$\div 15.8$	$- 1.1998$	
	$\overline{\overline{1}}.4791$	$17°30' = \angle AOC$

Vector addition has all the formal properties of addition of real numbers; in particular,

(1) $$\vec{u} + \vec{v} = \vec{v} + \vec{u},$$

(2) $$(\vec{u} + \vec{v}) + \vec{w} = \vec{u} + (\vec{v} + \vec{w}),$$

are true for any vectors $\vec{u}$, $\vec{v}$ and $\vec{w}$. Property (1) follows from Definition 3.1. Property (2) is obvious geometrically (see Figure 5.20). It follows that the resultant of three vectors $\vec{u}$, $\vec{v}$, $\vec{w}$ is well-defined without specifying the order of vectors, nor the order in which vector additions are performed. Thus we can write $\vec{u} + \vec{v} + \vec{w}$ instead of $(\vec{u} + \vec{v}) + \vec{w}$ or $\vec{u} + (\vec{v} + \vec{w})$. We can similarly show that the same is true about any number of vectors: their resultant does not depend on the order of the vectors nor on their groupings by parentheses.

Figure 5.20

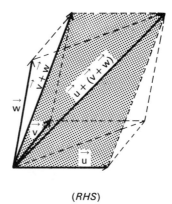

(LHS)   (RHS)

In the remainder of this section we shall give examples of applications of vectors in physics. We shall limit ourselves to 2-dimensional problems (i.e., in a plane) involving either forces or velocities. Forces and velocities are vectors; these facts follow from the laws of physics and geometry. In other words, if, for example, two forces act at a point, then the combined effect of these forces is that of their resultant, the force corresponding to the vector sum of the two forces as defined in Definition 3.1, part (ii).

Example 3.3 (a) A force $\vec{f}_1$ of 10 lbs. and a force $\vec{f}_2$ of 15 lbs. are acting on a spring balance that registers 13 lbs. What is the angle between the two forces? Since there is no motion, the two forces are in equilibrium with $\vec{f}_3$, the force exerted by the spring, i.e., $\vec{f}_1 + \vec{f}_2 + \vec{f}_3 = 0$. In other words, the resultant of $\vec{f}_1$ and $\vec{f}_2$ is equal to $-\vec{f}_3$. The forces $\vec{f}_1$ and $\vec{f}_2$ are represented in Figure 5.21 by $\overrightarrow{OA}$ and $\overrightarrow{OB}$, respectively.

Figure 5.21

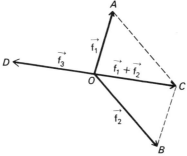

The problem is to evaluate $\angle AOB$, or equivalently, $\angle OAC = 180° - \angle AOB$. We apply the law of cosines to triangle OAC:

$$\cos(\angle OAC) = \frac{OA^2 + AC^2 - OC^2}{2OA \times AC}$$

$$= \frac{10^2 + 15^2 - 13^2}{2 \times 10 \times 15}$$

$$= \frac{156}{300}$$

$$= 0.52.$$

We obtain from Table 2,

$$\angle OAC = 58°30',$$

and therefore

$$\angle AOB = 121°30'.$$

(b) Two forces, $\vec{f_1}$ and $\vec{f_2}$, act at a point. If $f_1 = 18$ lbs., the angle between the forces is 42°, and the magnitude of the resultant is 23 lbs., find the magnitude of $\vec{f_2}$.

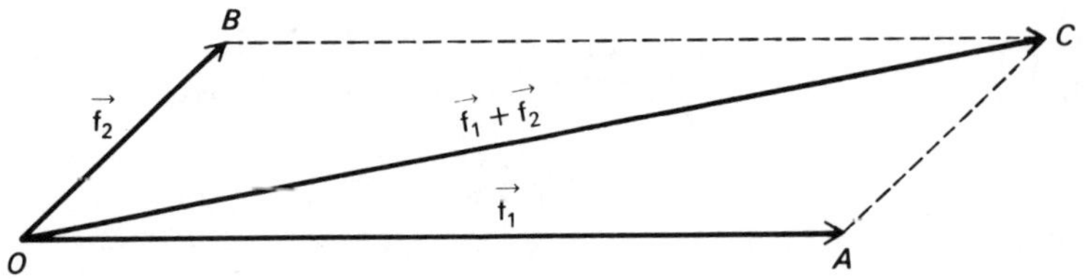

Figure 5.22

The forces are depicted in Figure 5.22. In triangle OAC, $OA = 18$, $OC = 23$, $\angle OAC = 180° - \angle AOB = 138°$. First we use the law of sines to compute

$$\sin(\angle OCA) = \frac{OA}{OC} \sin(\angle OAC)$$

$$= \frac{18}{23} \sin 138°.$$

Numbers	Logs	Antilogs
18	1.2553	
$\times \sin 138°$		
$(= \sin 42°)$	$+ \overline{1}.8255$	
	1.0808	
$\div 23$	$- 1.3617$	
	$\overline{1}.7191$	$31°30' = \angle OCA$

(Note that there is one solution only: $\angle OCA$ cannot be obtuse, since $\angle OAC$ is obtuse.)

Now, the required magnitude OB is equal to AC. Therefore, using the law of sines again,

$$AC = OA \frac{\sin (\angle AOC)}{\sin (\angle OCA)}$$

$$= 18 \frac{\sin (42° - 31°30')}{\sin 31°30'}.$$

Numbers	Logs	Antilogs
18	1.2553	
$\times$ sin 10°30'	$+ \bar{1}.2606$	
	0.5159	
$\div$ sin 31°30'	$- \bar{1}.7191$	
	0.7968	$6.3 = AC$

Hence $f_2 = 6.3$ lbs.

(c)   A plane is headed in the direction N40°W and is flying with an air speed of 200 m.p.h. If the wind is blowing from the South at 30 m.p.h., find the direction and ground speed of the plane. (Recall that the magnitude of velocity is called speed.)

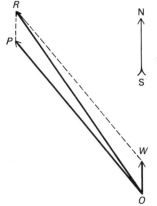

Figure 5.23

In Figure 5.23, the velocity of wind is represented by $\overrightarrow{OW}$, the air velocity of the plane by $\overrightarrow{OP}$, and the resultant, i.e., the ground velocity of the plane, by $\overrightarrow{OR}$. In triangle OPR, $OP = 200$, $PR = OW = 30$, $\angle OPR = 180° - \angle POW = 180° - 40° = 140°$. Using the law of tangents,

$$\frac{OP - PR}{OP + PR} = \frac{\tan \frac{1}{2}(\angle PRO - \angle POR)}{\tan \frac{1}{2}(180° - \angle OPR)},$$

i.e.,

$$\tan \frac{1}{2}(\angle PRO - \angle POR) = \tfrac{170}{230} \tan 20°.$$

Numbers	Logs	Antilogs
170	2.2304	
× tan 20°	+ 1.5611	
	1.7915	
÷ 230	− 2.3617	
	1.4298	$15° = \frac{1}{2}(\angle PRO - \angle POR)$

Thus

$$\tfrac{1}{2}\angle PRO - \tfrac{1}{2}\angle POR = 15°,$$
$$\tfrac{1}{2}\angle PRO + \tfrac{1}{2}\angle POR = 20°,$$

i.e.,

$$\angle PRO = 35°, \ \angle POR = 5°.$$

Hence the direction of the ground velocity is N35°W. Now we use the law of sines:

$$OR = OP \frac{\sin (\angle OPR)}{\sin (\angle PRO)}$$

$$= 200 \frac{\sin 140°}{\sin 35°}.$$

Numbers	Logs	Antilogs
200	2.3010	
× sin 140°		
(= sin 40°)	+ 1.8081	
	2.1091	
÷ sin 35°	− 1.7586	
	2.3505	224 = OR

The ground speed of the plane is 224 m.p.h.

(d) Wind coming from N75°W blows a raft in the direction S68°E. The raft is in a river flowing due west at the rate of 4 m.p.h. What is the speed of the raft relative to the ground?

The velocity of the raft is the resultant of the velocities of the wind and of the current. In Figure 5.24, vectors $\overrightarrow{OW}$, $\overrightarrow{OR}$, and $\overrightarrow{OC}$ represent the velocities of the wind, of the raft, and of the current, respectively. The length OC represents the speed of 4 m.p.h. We have to evaluate, to the same scale, the length OR. In the triangle ORW, $WR = OC = 4$, $\angle ROW = 75° - 68° = 7°$, $\angle OWR = 90° - 75° = 15°$. Therefore, using the law of sines,

Figure 5.24

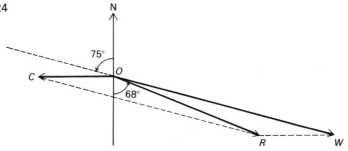

$$OR = WR \frac{\sin (\angle OWR)}{\sin (\angle ROW)}$$

$$= 4 \frac{\sin 15°}{\sin 7°} \cdot$$

Numbers	Logs	Antilogs
4	0.6021	
$\times \sin 15°$	$+ \bar{1}.4130$	
	0.0151	
$\div \sin 7°$	$- \bar{1}.0859$	
	0.9292	$8.5 = OR$

SYNOPSIS

In this section we introduced vectors. We discussed their geometric representation and gave examples of applications of trigonometry in solving problems involving forces and velocities.

QUIZ

Answer *true* or *false:*
1. $\vec{v} + (-\vec{v}) = \vec{0}$, for any vector $\vec{v}$.
2. If $\overrightarrow{OC} = \overrightarrow{OA} + \overrightarrow{OB}$, then $OC = OA + OB$.
3. If $\overrightarrow{OA} + \overrightarrow{OD} = \overrightarrow{OB} + \overrightarrow{OD}$, then $\overrightarrow{OA} = \overrightarrow{OB}$.
4. If $\overrightarrow{OP} + \overrightarrow{OR} = \overrightarrow{OS}$, then $OP \leq OS$.
5. If $2\vec{u} = \vec{u} + \vec{v}$, then $\vec{u} = \vec{v}$.
6. If $OA = OB$ and $\angle AOB = 90°$, then the magnitude of $\overrightarrow{OA} + \overrightarrow{OB}$ is $\sqrt{2}\, OA$.
7. If $OP = OR$ and the magnitude of $\overrightarrow{OP} + \overrightarrow{OR}$ is $\sqrt{2}\, OP$, then $\angle POR = 90°$.
8. If the magnitudes of $\vec{u}, \vec{v}, \vec{w}$ are positive then $\vec{u} + \vec{v} + \vec{w} \neq \vec{0}$.
9. $(\vec{u} + \vec{v}) + (\vec{r} + \vec{s}) = (\vec{u} + \vec{r}) + (\vec{v} + \vec{s})$, for any vectors $\vec{u}, \vec{v}, \vec{r}, \vec{s}$.
10. $c(\vec{u} + \vec{v}) = c\vec{u} + c\vec{v}$, for any vectors $\vec{u}$ and $\vec{v}$ and any number c.

EXERCISES

1. If $\overrightarrow{OC} = \overrightarrow{OA} + \overrightarrow{OB}$ and $OC = OA = OB$, find $\angle AOB$.

2. Let O, A, B be the points $(0,0)$, $(1,0)$, and $(0,1)$ relative to a rectangular system of coordinates. Let $\overrightarrow{OA} = \vec{i}$ and $\overrightarrow{OB} = \vec{j}$.
   (a) If $\overrightarrow{OC} = x\vec{i} + y\vec{j}$, where x and y are numbers, what are the coordinates of C?
   (b) Show that any given vector $\overrightarrow{OP}$ can be represented in the form $x\vec{i} + y\vec{j}$, where x and y are uniquely determined numbers. Show that Definition 3.1 implies that

$$(x_1\vec{i} + y_1\vec{j}) + (x_2\vec{i} + y_2\vec{j}) = (x_1 + x_2)\vec{i} + (y_1 + y_2)\vec{j}.$$

3. Find the magnitude and direction of the resultant of a force of 5 lbs. and a force of 10 lbs., if the forces act:
   (a) in the same direction;
   (b) in opposite directions;
   (c) at an angle of $60°$;
   (d) at an angle of $90°$;
   (e) at an angle of $120°$.

4. The direction of vectors $\overrightarrow{OA}$ and $\overrightarrow{OB}$ are N40°E and S50°E, respectively. If $OA = 2\,OB$, find the direction of $\overrightarrow{OA} + \overrightarrow{OB}$.

5. The magnitudes of forces $\vec{f}_1, \vec{f}_2, \vec{f}_3$ are 4 lbs., 6 lbs., 7 lbs., and their directions are N30°W, N80°W, and S30°W, respectively. Find the magnitude and the direction of the resultant force $\vec{f}_1 + \vec{f}_2 + \vec{f}_3$.

6. Three forces represented by vectors $\overrightarrow{OA}, \overrightarrow{OB}, \overrightarrow{OC}$ are in equilibrium. Show that

$$\frac{OA}{\sin \angle BOC} = \frac{OB}{\sin \angle COA} = \frac{OC}{\sin \angle AOB}.$$

7. Three forces $\vec{f}_1, \vec{f}_2, \vec{f}_3$ are in equilibrium. The angles between the forces are denoted by α, β, γ, as shown in Figure 5.25.

Figure 5.25

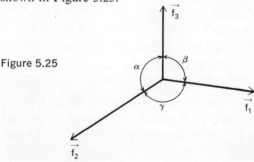

In each of the following exercises determine the unknown forces and angles:
(a) $f_1 = 12$ lbs., $f_2 = 15$ lbs., $f_3 = 11$ lbs.;
(b) $f_1 = 22$ lbs., $f_2 = 19$ lbs., $\gamma = 125°$;
(c) $f_1 = 5.65$ lbs., $\alpha = 110°$, $\beta = 135°$;
(d) $f_1 = f_2 = 17$ lbs., $\gamma = 55°$.

8. A boat travels at 8 m.p.h. across a river flowing 3 m.p.h. In what direction should the boat be headed so as to reach the point exactly opposite the starting point?

9. A kite is inclined 60° to the vertical. It remains in equilibrium under its weight and the pull of a cord inclined 50° to the horizontal. If the tension in the cord is 4.5 lbs., what is the weight of the kite?

10. A weight of 35 lbs., suspended by a rope, is pulled aside by a rope which makes an angle of 12° with the horizontal in such a way that the first rope makes an angle of 25° with the vertical. What are the tensions in the ropes?

11. A school bus is moving on a straight road at a speed of 30 m.p.h. Inside the bus boy X throws a ball to boy Y at horizontal speed of 12 m.p.h. The line XY makes an angle of 38° with the direction of the road.
(a) What is the horizontal velocity of the ball relative to the road?
(b) If Y returns the ball at the same speed, what is its velocity relative to the road?

12. Relative to the ground a plane is flying 272 m.p.h. in the S35°W direction. It is headed in the S42°W direction and its air speed is 295 m.p.h. What is the velocity of the wind?

13. A train is moving northward with a speed of 15 m.p.h. A fly flies inside the train with a speed of 7 m.p.h. so that its velocity relative to the ground is N20°E. Find the speed of the fly relative to the ground. [*Note:* There are two solutions.]

Circular Functions

6.1 Circular Functions

Trigonometry, which is one of the oldest branches of mathematics, was originally just a study of the relations between the sides and the angles of triangles. Even today trigonometry of this type has extensive applications in experimental physics, engineering, and surveying, and, in the form of spherical trigonometry (the treatment of triangles on the surface of a sphere), it also plays a major role in navigation and astronomy. It is rather surprising that, apart from these direct applications, trigonometric functions (in a somewhat abstract guise) play an important part in pure mathematics: in the differential and integral calculus and in other parts of mathematical analysis.

The abstraction of trigonometric functions follows the same pattern as the abstraction of the concept of numbers. Trigonometric functions as defined in Sections 3.1 and 3.2 have the set of angles as their domain. This limits their possible applications in mathematics because angles are specialized geometric concepts and occur only in rather limited parts of this field. Of course, the value of a trigonometric function at any angle depends only on the measure of the angle and not on its position in the plane, in space, nor on any other geometric characteristic. In other words, once a unit of angle measure has been chosen, the domain of a trigonometric function can be regarded as a set of real numbers. This idea is developed in the following definition.

Let R denote the set of real numbers, and for any $x \in R$, let α_x denote an angle of x radians. Let

$$R_1 = \{x \mid x \in R, x \neq (2k + 1)\pi/2, k \in Z\}$$

and

$$R_2 = \{x \mid x \in R, x \neq k\pi, k \in Z\},$$

i.e., let R_1 be the set of all real numbers that are not odd multiples of $\pi/2$ and let R_2 be the set of all real numbers that are not multiples of π.

Definition 1.1 (Circular Sine, Cosine, and Tangent) *The circular sine function* $S\colon R \to R$ is defined by

$$S(x) = \sin \alpha_x$$

for all $x \in R$. Similarly, the *circular cosine function* $C\colon R \to R$ is defined by

$$C(x) = \cos \alpha_x$$

for all $x \in R$, and the *circular tangent function* $T\colon R_1 \to R$ by

$$T(x) = \tan \alpha_x$$

for all $x \in R_1$.

The circular cosecant, secant, and cotangent functions are likewise defined on R_2 to R, R_1 to R, and R_3 to R, respectively. In general, *the value of a circular function at a real number x is the value of the associated trigonometric function at an angle of x radians.* It is customary to use the same notation for circular functions as for the corresponding trigonometric functions. For example, we write

$$\sin \pi = 0, \tan \frac{\pi}{4} = 1, \text{ etc.}$$

Example 1.1 Interpret geometrically the meaning of circular functions without reference to angles.

Let U denote the set of points on the unit circle. Let $\varphi\colon R \to U$ be a function defined as follows: For any nonnegative $t \in R$ the value $\varphi(t)$ is the point on the unit circle whose distance from $(1, 0)$, measured counterclockwise along the circle, is t units (Figure 6.1).

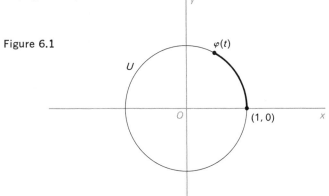

Figure 6.1

If t is negative, the point $\varphi(t)$ is determined similarly except that the distance of $-t$ units from $(1, 0)$ is measured clockwise. For example, the positions of points $\varphi(\pi/4)$, $\varphi(-\pi/4)$, $\varphi(2\pi/3)$, $\varphi(3\pi/4)$, $\varphi(\pi)$, $\varphi(-8\pi/3)$, $\varphi(11\pi/4)$, and $\varphi(5\pi)$ are shown in Figure 6.2.

Figure 6.2

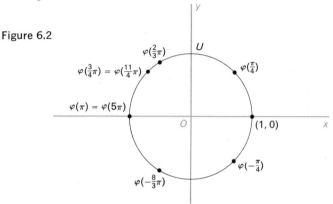

Note that $\varphi(3\pi/4) = \varphi(11\pi/4)$, $\varphi(5\pi) = \varphi(\pi)$, etc. In fact, since the circumference of the unit circle is 2π units, the function φ must repeat itself exactly after every interval of length 2π, i.e.,

(1) $$\varphi(t + 2k\pi) = \varphi(t)$$

for any $t \in R$ and any $k \in Z$. We say that φ is *periodic* with *period* 2π. Now, it follows from the definitions of circular functions that if the coordinates of $\varphi(t)$ are (x, y), then these functions can also be given by:

$$\sin t = y,$$

$$\cos t = x,$$

$$\tan t = \frac{y}{x}, \quad \text{if} \quad x \neq 0,$$

$$\operatorname{cosec} t = \frac{1}{y}, \quad \text{if} \quad y \neq 0,$$

$$\sec t = \frac{1}{x}, \quad \text{if} \quad x \neq 0,$$

$$\cot t = \frac{x}{y}, \quad \text{if} \quad y \neq 0.$$

Note that (1) and the above definitions imply that all these circular functions are periodic with period not exceeding 2π. This conclusion, of course, also follows directly from Definition 1.1.

The definitions of circular functions imply that all the formulas and identities established in Sections 3.2 and 4.1 for trigonometric functions have exact analogues for the corresponding circular functions. We list some of them here for reference: For any x, y in R, for which the functions are defined,

(2) $$\sin(-x) = -\sin x,$$

(3) $$\cos(-x) = \cos x,$$

(4) $$\tan(-x) = -\tan x,$$

(5) $$\sin(\pi - x) = \sin x,$$

(6) $$\cos(\pi - x) = -\cos x,$$

(7) $$\tan(\pi - x) = -\tan x,$$

(8) $$\sin\left(\frac{\pi}{2} - x\right) = \cos x,$$

(9) $$\cos\left(\frac{\pi}{2} - x\right) = \sin x,$$

(10) $$\tan\left(\frac{\pi}{2} - x\right) = \cot x.$$

We defined circular functions in order to introduce functions whose properties and behavior mimic those of the corresponding trigonometric functions, but whose domains and ranges are sets of real numbers. This allows us to study the properties of a circular function by analytical rather than geometric means and, in particular, to construct its graph and thereby visualize its behavior.

Example 1.2 (a) Use the values of trigonometric functions in Table 2 to determine, via Definition 1.1, the values of circular functions for all integer multiples of $\pi/12$ between 0 and $\pi/2$. Use these data to sketch the graphs of these functions in the interval $[-2\pi, 2\pi]$.

x	0	$\pi/12$ $=0.2618$	$\pi/6$ $=0.5236$	$\pi/4$ $=0.7854$	$\pi/3$ $=1.047$	$5\pi/12$ $=1.309$	$\pi/2$ $=1.571$
$\sin x$	.0000	.2588	.5000	.7071	.8660	.9659	1.000
$\cos x$	1.000	.9659	.8660	.7071	.5000	.2588	.0000
$\tan x$	.0000	.2679	.5774	1.000	1.732	3.732	undefined

The other required values in the interval $[-2\pi, 2\pi]$ are easily computed from trigonometric reduction formulas and from the periodicity of circular functions.

Figure 6.3

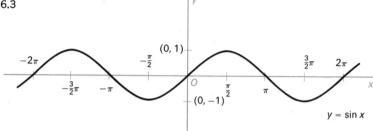

$y = \sin x$

Figure 6.4

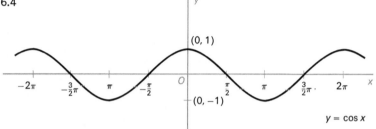

$y = \cos x$

Figure 6.5

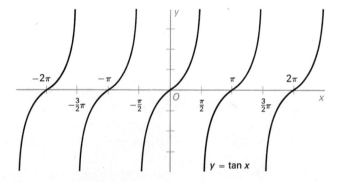

$y = \tan x$

(b) Compute the values of the cosecant, secant, and cotangent functions for all integer multiples of $\pi/12$ between 0 and $\pi/2$. Use these data to sketch the graphs of these functions in the interval $[-2\pi, 2\pi]$.

x	0	$\pi/12$	$\pi/6$	$\pi/4$	$\pi/3$	$5\pi/12$	$\pi/2$
cosec x	undefined	3.864	2.000	1.414	1.155	1.035	1.000
sec x	1.000	1.035	1.155	1.414	2.000	3.864	undefined
cot x	undefined	3.732	1.732	1.000	.5774	.2679	.0000

Figure 6.6

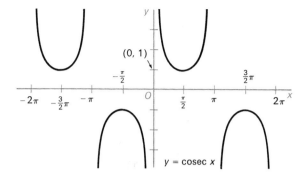

$y = \text{cosec } x$

Figure 6.7

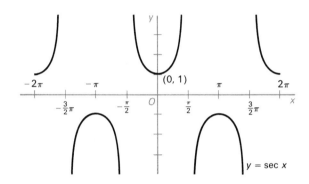

$y = \text{sec } x$

Figure 6.8

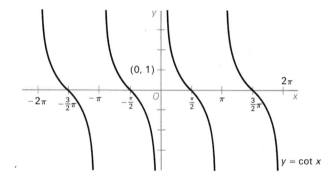

$y = \text{cot } x$

Example 1.3 Show that

(11)
$$\sin a < a < \tan a$$

for any number a, $0 < a < \dfrac{\pi}{2}$.

Let α be the angle of a radians (Figure 6.9 on the next page), and take $OP = OR = 1$. Recall (see Definitions 2.2 and 2.3, Chapter 3) that

Figure 6.9

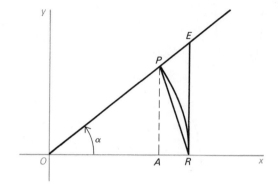

$$\sin a = \sin \alpha = AP,$$
$$\tan a = \tan \alpha = RE.$$

Denote by $\triangle ORP$ and $\triangle ORE$ the areas of triangles ORP and ORE, and by $\sigma(ORP)$ the area of the sector ORP of the unit circle. We have from the geometry of the figure

(12) $$\triangle ORP < \sigma(ORP) < \triangle ORE.$$

Now, the ratio of $\sigma(ORP)$ to the area of the whole circle is equal to the ratio of angle α to a complete revolution, i.e.,

$$\frac{\sigma(ORP)}{\pi \times 1^2} = \frac{\alpha}{2 \times 1 \times \pi}$$

and therefore

(13) $$\sigma(ORP) = \tfrac{1}{2}\alpha.$$

Also,

$$\triangle ORP = \tfrac{1}{2}OR \times AP$$
(14) $$= \tfrac{1}{2} \times 1 \times \sin \alpha$$
$$= \tfrac{1}{2} \sin \alpha,$$

and

$$\triangle ORE = \tfrac{1}{2}OR \times RE$$
(15) $$= \tfrac{1}{2} \times 1 \times \tan \alpha$$
$$= \tfrac{1}{2} \tan \alpha.$$

Hence from (12), (14), (13), and (15),

$$\tfrac{1}{2} \sin \alpha < \tfrac{1}{2}\alpha < \tfrac{1}{2} \tan \alpha$$

and therefore

$$\sin a < a < \tan a.$$

SYNOPSIS

In this section we defined circular functions in terms of corresponding trigonometric functions and discussed their properties and their graphs. We also established the inequalities

$$\sin a < a < \tan a,$$

for any number a, $0 < a < \pi/2$.

QUIZ

Answer *true* or *false:*
1. The equation $\sec x = \frac{1}{2}$ has no solutions.
2. If φ is the function given in Example 1.1, then $\varphi(\pi) = \varphi(-\pi)$.
3. If φ is the same function as above, then $\varphi(0) = (0, 0)$.
4. $\sin (x - \pi) = \sin (x + \pi)$ for any number x.
5. $\sin x = \tan x$ if and only if $\sin x = 0$.
6. $\tan (\pi + x) = \tan x$, for any $x \in R$ for which $\tan x$ is defined.
7. $\sin (\frac{3}{2}\pi - x) = -\cos x$, for any $x \in R$.
8. $\sin (2\pi - x) = \sin x$, for all $x \in R$.
9. If the indicated functions are understood to be circular functions, then $\sin (\cos \pi/2) = 0$.
10. $\cos (\sin \pi/2) = 0$.

EXERCISES

1. Show that

$$1 < \frac{x}{\sin x} < \sec x,$$

for any positive x smaller than $\pi/2$.
2. A function is called *even* or *odd* according as $f(-x) = f(x)$ or $f(-x) = -f(x)$, for all x in its domain. Which of the circular functions are even and which are odd? Is every real-valued function either even or odd?
3. A function is said to be *increasing* in an interval if $f(x_1) < f(x_2)$ whenever $x_1 < x_2$ in the interval. It is called *decreasing* if $f(x_1) > f(x_2)$ whenever $x_1 < x_2$. Determine the intervals between 0 and 2π in which each of the circular functions increases and the intervals in which each decreases.

4.   Sketch graphs of each of the following functions from -2π to 2π:

(a) $f(x) = 1 + \sin x$;

(b) $f(x) = \sin x - 1$;

(c) $f(x) = -\sin x$;

(d) $f(x) = 1 - \sin x$;

(e) $f(x) = 1 + \cos x$;

(f) $f(x) = 1 + \tan x$;

(g) $f(x) = -\tan x$;

(h) $f(x) = 1 - \dfrac{\cos^2 x}{1 + \sin x}$.

[*Hint:* Simplify the right-hand side of the formula.]

6.2   Inverse Circular Functions

Inverse circular functions can be defined in the same way as were the inverse trigonometric functions in Section 3.4.   None of the circular functions defined in the preceding sections is 1–1.   If, however, their domains and ranges are suitably restricted, the resulting functions are 1–1 onto and possess inverses.

Definition 2.1   (Inverse Circular Functions)   The *inverse circular sine* function $\sin^{-1}$: $[-1, 1] \rightarrow \left[-\dfrac{\pi}{2}, \dfrac{\pi}{2}\right]$ is a real-valued function whose defining relation is

(1) $\sin (\sin^{-1} x) = x$.

Similarly, we define the other five inverse circular functions: The *inverse cosine* function $\cos^{-1}$: $[-1, 1] \rightarrow [0, \pi]$ is defined by

(2) $\cos (\cos^{-1} x) = x$.

The *inverse tangent* function $\tan^{-1}: R \to \left(-\dfrac{\pi}{2}, \dfrac{\pi}{2} \right)$ is defined by

(3) $$\tan (\tan^{-1} x) = x.$$

The *inverse cotangent* function $\cot^{-1}: R \to (0, \pi)$ is defined by

(4) $$\cot (\cot^{-1} x) = x.$$

The *inverse cosecant* function $\operatorname{cosec}^{-1}: X \to Y_1$, where $X = \{x \mid x \leq -1$ or $x \geq 1\}$ and $Y_1 = \left\{ y \mid -\dfrac{\pi}{2} \leq y < 0 \text{ or } 0 < y \leq \dfrac{\pi}{2} \right\}$, is defined by

(5) $$\operatorname{cosec} (\operatorname{cosec}^{-1} x) = x.$$

The *inverse secant* function $\sec^{-1}: X \to Y_2$, where X is the set defined above and $Y_2 = \left\{ y \mid 0 \leq y < \dfrac{\pi}{2} \text{ or } \dfrac{\pi}{2} < y \leq \pi \right\}$, is defined by

(6) $$\sec (\sec^{-1} x) = x.$$

Theorem 2.1

(a) *If* $-1 \leq x \leq 1$, *then* $\cos^{-1} x = \dfrac{\pi}{2} - \sin^{-1} x.$

(b) *For any* $x \in R$, $\cot^{-1} x = \dfrac{\pi}{2} - \tan^{-1} x.$

(c) *If* $x \leq -1$ *or* $x \geq 1$, $\operatorname{cosec}^{-1} x = \sin^{-1} \dfrac{1}{x}.$

(d) *If* $x \leq -1$ *or* $x \geq 1$, $\sec^{-1} x = \dfrac{\pi}{2} - \operatorname{cosec}^{-1} x.$

Proof (a) Set
$$t = \sin^{-1} x.$$

Then
$$x = \sin t$$

(7) $$= \cos \left(\dfrac{\pi}{2} - t \right).$$

From (7) we cannot conclude immediately that $\cos^{-1} x = \dfrac{\pi}{2} - t = \dfrac{\pi}{2} - \sin^{-1} x$.

We first have to show that for all values of x, the number $\dfrac{\pi}{2} - t$ lies in the range

of $\cos^{-1}$. Now if $-1 \le x < 0$, then $-\dfrac{\pi}{2} \le \sin^{-1} x \le 0$, and therefore

$$\frac{\pi}{2} \le \frac{\pi}{2} - \sin^{-1} x \le \pi,$$

i.e.,

$$\frac{\pi}{2} \le \frac{\pi}{2} - t \le \pi.$$

Hence in this case $\dfrac{\pi}{2} - t$ belongs to the range of $\cos^{-1}$. If $0 \le x \le 1$, then

$0 \le \sin^{-1} x \le \dfrac{\pi}{2}$, and therefore

$$0 \le \frac{\pi}{2} - \sin^{-1} x \le \frac{\pi}{2},$$

i.e.,

$$0 \le \frac{\pi}{2} - t \le \frac{\pi}{2},$$

and again $\dfrac{\pi}{2} - t$ lies in the range of $\cos^{-1}$. Thus (7) implies that

$$\cos^{-1} x = \frac{\pi}{2} - t$$

$$= \frac{\pi}{2} - \sin^{-1} x.$$

(b) Set

$$t = \tan^{-1} x.$$

Then

$$x = \tan t$$

$$= \cot\left(\frac{\pi}{2} - t\right).$$

We can show, as in (a), that $\dfrac{\pi}{2} - t$ belongs to the range of $\cot^{-1}$ and therefore

$$\cot^{-1} x = \frac{\pi}{2} - t$$

$$= \frac{\pi}{2} - \tan^{-1} x.$$

(c)   Here we set

$$t = \sin^{-1} \frac{1}{x}$$

and obtain

$$\sin t = \frac{1}{x},$$

i.e.,

(8) $$\cosec t = x.$$

Now, t lies in the range of $\sin^{-1}$, i.e., in $\left[-\frac{\pi}{2}, \frac{\pi}{2} \right]$ and $t \neq 0$, since $\frac{1}{x} \neq 0$. Thus t is in the range of $\cosec^{-1}$, and therefore (8) implies that

$$t = \cosec^{-1} x,$$

i.e.,

$$\cosec^{-1} x = \sin^{-1} \frac{1}{x}.$$

(d)   This identity may be proved by a method analogous to the one used to prove the identity in (a). ∎

The graph of a function $f: X \to Y$, where X and Y are subsets of R, is the set of points $\{(\alpha, \beta) \mid \beta = f(\alpha), \alpha \in X\}$. If f happens to be 1–1 onto Y, then f has an inverse $f^{-1}: Y \to X$. The graph of the inverse function is the set of points $\{(\beta, \alpha) \mid \alpha = f^{-1}(\beta), \beta \in Y\} = \{(\beta, \alpha) \mid \beta = f(\alpha), \alpha \in X\}$, since f is 1–1 onto Y. It follows that (α, β) is a point of the graph of f if and only if (β, α) is a point of the graph of f^{-1}. In other words, the graphs of f and of f^{-1} are reflections of each other in the line $y = x$.

Example 2.1   Sketch the graphs of the inverse circular functions $\sin^{-1}$, $\cos^{-1}$, $\tan^{-1}$.

In the preceding section we sketched the graph of the sine function (Figure 6.3). The function whose inverse is $\sin^{-1}$ is the sine function restricted to the domain $\left[-\frac{\pi}{2}, \frac{\pi}{2} \right]$. In Figure 6.10 the graphs of the functions sin: $\left[-\frac{\pi}{2}, \frac{\pi}{2} \right] \to [-1, 1]$

Figure 6.10

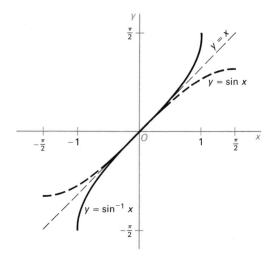

and $\sin^{-1}$: $[-1, 1] \rightarrow \left[-\dfrac{\pi}{2}, \dfrac{\pi}{2}\right]$ are drawn in a broken line and solid line, re-
specitively; they are symmetric relative to the line $y = x$. In a similar manner we
sketch the graphs of the two other inverse circular functions (Figures 6.11 and 6.12).
In each figure the direct function, restricted to a suitable domain, is drawn in a
broken line, and the inverse function in a solid line.

Figure 6.11

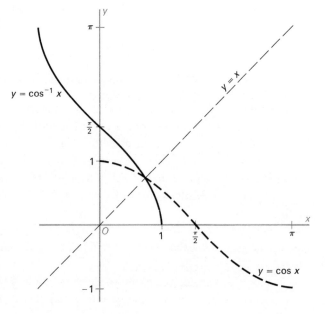

Figure 6.12

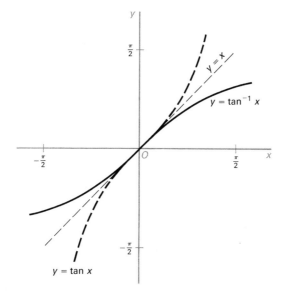

Example 2.2   Show that if x and y are real numbers satisfying

(9) $$x^2 + y^2 = 1,$$

then there exists a real number t such that $x = \cos t$ and $y = \sin t$.
   The equation (9) implies that $x^2 \leq 1$ and $y^2 \leq 1$ and therefore that

$$-1 \leq x \leq 1 \text{ and } -1 \leq y \leq 1.$$

It follows that x lies in the domain of $\cos^{-1}$. Suppose first that y is nonnegative, and set $t = \cos^{-1} x$. Then $0 \leq t \leq \pi$, $x = \cos t$, and

$$
\begin{aligned}
y &= \sqrt{1 - x^2} \\
&= \sqrt{1 - \cos^2 t} \\
&= \sqrt{\sin^2 t} \\
&= \sin t,
\end{aligned}
$$

since $\sin t \geq 0$. If $y < 0$, set $t = 2\pi - \cos^{-1} x$. Then $\pi \leq t < 2\pi$, and

$$
\begin{aligned}
\cos t &= \cos (2\pi - \cos^{-1} x) \\
&= \cos (\cos^{-1} x) \\
&= x.
\end{aligned}
$$

(Recall that $\cos(2\pi - z) = \cos z$ for all z.)   Also,

$$
\begin{aligned}
y &= -\sqrt{1 - x^2} \\
&= -\sqrt{1 - \cos^2 t} \\
&= -\sqrt{\sin^2 t} \\
&= \sin t,
\end{aligned}
$$

since $\pi \le t < 2\pi$ and therefore $\sin t \le 0$.

We now prove some standard identities for inverse circular functions.

Theorem 2.2   *The following identities hold for any number x in the indicated sets of real numbers:*

(a) $\qquad \sin^{-1}(-x) = -\sin^{-1} x, \quad if -1 \le x \le 1;$

(b) $\qquad \cos^{-1}(-x) = \pi - \cos^{-1} x, \quad if -1 \le x \le 1;$

(c) $\qquad \tan^{-1}(-x) = -\tan^{-1} x, \quad for\ all\ x \in R;$

(d) $\qquad \sin^{-1} x = \cos^{-1}\sqrt{1 - x^2}, \quad if\ 0 \le x \le 1;$

(e) $\qquad \sin^{-1} x = \tan^{-1}\dfrac{x}{\sqrt{1 - x^2}}, \quad if -1 < x < 1;$

(f) $\qquad \tan^{-1} x = \sin^{-1}\dfrac{x}{\sqrt{1 + x^2}}, \quad for\ all\ x \in R;$

(g) $\qquad \tan^{-1} x = \cos^{-1}\dfrac{1}{\sqrt{1 + x^2}}, \quad if\ x \ge 0.$

Proof   The technique of proof is similar for all these identities.   We shall prove three of them and leave the proofs of the other equalities as an exercise for the student.

(a)   Set $t = \sin^{-1} x$, then

$$x = \sin t,$$

and therefore

$$
\begin{aligned}
\sin^{-1}(-x) &= \sin^{-1}(-\sin t) \\
&= \sin^{-1}(\sin(-t)).
\end{aligned}
$$

Now,

$$-\frac{\pi}{2} \leq \sin^{-1} x \leq \frac{\pi}{2},$$

i.e.,

$$-\frac{\pi}{2} \leq t \leq \frac{\pi}{2},$$

or

$$-\frac{\pi}{2} \leq -t \leq \frac{\pi}{2}.$$

Thus $-t$ belongs to the range of $\sin^{-1}$, and therefore

$$\sin^{-1}\left(\sin\left(-t\right)\right) = -t$$
$$= -\sin^{-1} x,$$

and the identity is proved.

(b)   We start as in the proof of (a); here we set $t = \cos^{-1} x$, i.e., $x = \cos t$ and obtain

$$\cos^{-1}\left(-x\right) = \cos^{-1}\left(-\cos t\right)$$
$$= \cos^{-1}\left(\cos\left(\pi - t\right)\right).$$

It is easy to verify that $\pi - t$ belongs to the range of $\cos^{-1}$, and therefore

$$\cos^{-1}\left(\cos\left(\pi - t\right)\right) = \pi - t$$
$$= \pi - \cos^{-1} x$$

and the result follows.

(f)   Let $t = \tan^{-1} x$, i.e., $x = \tan t$.   We compute

$$\sin^2 t = 1 - \cos^2 t$$
$$= 1 - \frac{1}{\sec^2 t}$$

(10)

$$= 1 - \frac{1}{1 + \tan^2 t}$$
$$= 1 - \frac{1}{1 + x^2}$$
$$= \frac{x^2}{1 + x^2}.$$

Now, $\sin t \geq 0$ if and only if $t \geq 0$, i.e., if and only if $\tan t \geq 0$ which occurs if and only if $x \geq 0$.   Similarly, $\sin t \leq 0$ if and only if $x \leq 0$.   Thus (10) implies that

$$\sin t = \frac{x}{\sqrt{1 + x^2}},$$

and since

$$-\frac{\pi}{2} < t < \frac{\pi}{2},$$

we have

$$t = \sin^{-1} \frac{x}{\sqrt{1 + x^2}},$$

and therefore

$$\tan^{-1} x = \sin^{-1} \frac{x}{\sqrt{1 + x^2}}. \ \blacksquare$$

Example 2.3 Evaluate $\cos\left((1/2) \sin^{-1} \frac{3}{5}\right)$.

Let $(1/2) \sin^{-1} \frac{3}{5} = t$. Then $\sin^{-1} \frac{3}{5} = 2t$, i.e., $\sin 2t = \frac{3}{5}$. We compute that

$$\frac{3}{5} = 2 \sin t \cos t$$
$$= 2\sqrt{1 - \cos^2 t} \cos t$$

since $0 < t < \pi/2$, and thus $0 < \sin t < 1$. It follows that

$$9/100 = (1 - \cos^2 t) \cos^2 t,$$

i.e.,

$$(\cos^2 t - 9/10)(\cos^2 t - 1/10) = 0.$$

Thus, remembering that $0 < t < \pi/2$, we get $\cos\left((1/2) \sin^{-1} \frac{3}{5}\right) = \cos t = \dfrac{3}{\sqrt{10}}$.

We ignore the other alternative, $\cos t = \dfrac{1}{\sqrt{10}}$, since $0 < t = (1/2) \sin^{-1} \frac{3}{5} < \pi/4$

while $\pi/4 < \cos^{-1} \dfrac{1}{\sqrt{10}} < \pi/2$.

SYNOPSIS

In this section we defined the inverse circular functions, sketched the graphs of $\sin^{-1}$, $\cos^{-1}$, and $\tan^{-1}$, and proved standard identities involving the inverse circular functions.

QUIZ

Answer *true* or *false:*

1. $\sin^{-1} x + \cos^{-1} x = \dfrac{\pi}{2}$, for all $x \in [-1, 1]$.

2. $\cot^{-1} x = 1/\tan^{-1} x$, for all x for which both sides are defined.
3. $\sec^{-1} x = 1/\cos^{-1} x$, for all x for which both sides are defined.
4. $\sin^{-1}(\sin x) = x$, for all x.
5. $\sin(\sin^{-1} x) = x$, for all x for which the left-hand side is defined.
6. If x belongs to the range of $\cos^{-1}$, then x is nonnegative.
7. $\pi - \sin^{-1} x = \sin^{-1} x$, for all x in the range of $\sin^{-1}$.

8. If $\sin^{-1} x = \cos^{-1} x$, then $x = \dfrac{1}{\sqrt{2}}$.

9. If $\sin^{-1} x = \frac{3}{5}$, then $\cos^{-1} x = \frac{4}{5}$.
10. If $\tan^{-1} x = 1$, then $\cot^{-1} x = 1$.

EXERCISES

1. Sketch the graphs of the inverse circular functions $\cot^{-1}$, $\operatorname{cosec}^{-1}$, and $\sec^{-1}$.
2. Prove part (d) of Theorem 2.1.
3. Prove parts (c), (d), (e), and (g) of Theorem 2.2.
4. Prove the identities:

   (a) $\operatorname{cosec}^{-1} x + \sec^{-1} x = \dfrac{\pi}{2}$;

   (b) $\cot^{-1} x = \tan^{-1} \dfrac{1}{x}$, whenever $x > 0$;

   (c) $\sin(\cos^{-1} x) = \cos(\sin^{-1} x)$.

5. The domains of $\sin^{-1}$ and $\cos^{-1}$ are both $[-1, 1]$. The range of $\sin^{-1}$ is $\left[-\dfrac{\pi}{2}, \dfrac{\pi}{2} \right]$. Can we alter the definition of $\cos^{-1}$, so that we would still have $\cos(\cos^{-1} x) = x$ for all $x \in [-1, 1]$, but the range of the new inverse cosine function would be $\left[-\dfrac{\pi}{2}, \dfrac{\pi}{2} \right]$? Why?

6. Prove or disprove the identities:
   (a) $\sin^{-1}(\sin x) = x$ for all x;
   (b) $\tan^{-1}(\tan x) = x$ for all x for which $\tan x$ is defined.

7. Find the explicit value of each of the following:
   (a) $\sin(2 \tan^{-1} \frac{1}{2})$
      [*Hint:* Recall that $\sin 2x = 2 \sin x \cos x$. Set $x = \tan^{-1} \frac{1}{2}$.];

(b) $\tan (2 \sin^{-1} \frac{1}{2})$;

(c) $\cos (2 \tan^{-1} \frac{1}{2})$;

(d) $\tan (2 \sin^{-1} \frac{2}{3})$;

(e) $\sin (\cos^{-1} \frac{1}{3} + \sin^{-1} \frac{1}{3})$

   [*Hint:* Recall that $\sin (\alpha + \beta) = \sin \alpha \cos \beta + \cos \alpha \sin \beta$.

   Set $\alpha = \cos^{-1} \frac{1}{3}$, $\beta = \sin^{-1} \frac{1}{3}$.];

(f) $\tan (\tan^{-1} \frac{2}{5} + \tan^{-1} \frac{3}{7})$;

(g) $\sin (\cos^{-1} \frac{13}{14} + \cos^{-1} \frac{11}{14})$;

(h) $\sin (\cos^{-1} \frac{3}{5} + \tan^{-1} \frac{4}{3})$;

(i) $\sin (\cos^{-1} \frac{1}{3} + \tan^{-1} \frac{1}{3})$;

(j) $\cos (\tan^{-1} 2 + \tan^{-1} 3)$.

8. Find an expression not involving circular functions nor their inverses for each of the following:

   (a) $\cos \left(\sin^{-1} \dfrac{x}{\sqrt{1 + x^2}} + \sin^{-1} \dfrac{1}{\sqrt{1 + x^2}} \right)$;

   (b) $\sin \left(\sin^{-1} \dfrac{1}{\sqrt{1 + x^2}} + \cos^{-1} \dfrac{x}{\sqrt{1 + x^2}} \right)$;

   (c) $\sin (\frac{1}{2} \cos^{-1} x)$;

   (d) $\cos (\frac{1}{2} \cos^{-1} x)$.

9. Prove that the number t in Example 2.2 is uniquely determined in the interval $0 \le t < 2\pi$.

6.3 Periodic Functions

   In the preceding sections we noted that the circular functions are periodic. In general if X and Y are sets of numbers, then a function $f: X \to Y$ is said to be *periodic* if there exists a nonzero element h in X such that

(1) $$f(x \pm h) = f(x),$$

for all x in X. Equation (1) implies that for any x in X both $x - h$ and $x + h$ must also be in X. The least positive h for which (1) holds is called the *period* of f. If f is a periodic function with period h and it has a graph, then the graph will repeat itself in intervals of h units along the x-axis.

Example 3.1   (a)   If x is any real number, let $[x]$ denote the largest integer not exceeding x, i.e.,

(2) $$x - 1 < [x] \le x.$$

For example, $[3.14] = 3$, $[3] = 3$, $[-3] = -3$, $[-3.14] = -4$, $[\sqrt{2}] = 1$, etc.

Let $f: R \rightarrow R$ be defined by

(3) $$f(x) = x - [x],$$

for all $x \in R$. Note that by (2),

$$[x] \leq x < [x] + 1,$$

and therefore

$$0 \leq x - [x] < 1.$$

We have, for example, $f(3.14) = 0.14$, $f(3) = 0$, $f(0.14) = 0.14$, $f(-3.14) = 0.86$, etc. In fact, if x is a nonnegative number written out as a decimal, then $f(x)$ is the fractional part of x, i.e., it is obtained from x by deleting the digits to the left of the decimal point. The graph of f is shown in Figure 6.13.

Figure 6.13

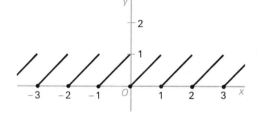

The graph consists of line segments each of which includes the left endpoint but not the right one. Clearly f is periodic with period 1.

(b) Let $g: R \rightarrow R$ be defined by

$$g(x) = \tfrac{3}{2}(-1)^{[x]}$$

The value of $g(x)$ is $\frac{3}{2}$ or $-\frac{3}{2}$ according as $[x]$ is even or odd. The graph of g is shown in Figure 6.14.

Figure 6.14

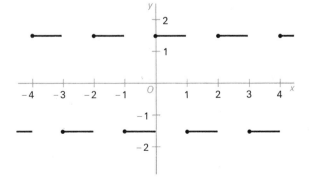

The function g is periodic with period 2.

Theorem 3.1   *The circular functions sine, cosine, cosecant, and secant are periodic with period 2π. The circular functions tangent and cotangent are periodic with period π.*

Proof   We noted in Example 1.1, Section 6.1, that if f is any of the six circular functions, then

$$f(x + 2k\pi) = f(x),$$

for any x in the domain of the function and any integer k. Hence all the six circular functions are periodic. We first show that the period of cosine is 2π, i.e., that if $0 < k < 2\pi$, then $\cos(x + k) \neq \cos x$ for some x. In fact, if we set $x = 0$ then $\cos x = 1$, while $\cos(x + k) = \cos k$, which cannot be 1 if $0 < k < 2\pi$. Thus 2π is the period of the cosine function. This implies immediately that 2π is also the period of the sine, cosecant, and secant functions. For, if h is a positive number such that

(4) $\sin(x + h) = \sin x$

for all x, then setting $x = \dfrac{\pi}{2} + z$, we have

$$\sin\left(\frac{\pi}{2} + z + h\right) = \sin\left(\frac{\pi}{2} + z\right),$$

i.e.,

(5) $\cos(z + h) = \cos z$

for all z. Since 2π is the period of the cosine function, equation (5) implies that $h \geq 2\pi$, and therefore that the least number h for which (4) holds cannot be less than 2π. It follows that 2π is the period of the sine function. It is easy to see that 2π must also be the period of the cosecant and secant functions (see Exercise 1).

Next we show that π is the period of the tangent function. For any x in domain of the tangent function we have

$$\tan(x + \pi) = \frac{\sin(x + \pi)}{\cos(x + \pi)}$$

$$= \frac{-\sin x}{-\cos x}$$

$$= \tan x.$$

It remains to prove that if $0 < h < \pi$, then $\tan(x + h) \neq \tan x$ for some x. Set $x = 0$. Then $\tan x = 0$, while $\tan(x + h) = \tan h$. Clearly $\tan h$ cannot be 0 if $0 < h < \pi$. We have proved that π is the period of the tangent function. It follows easily that π is the period of the cotangent function as well (see Exercise 2). ∎

In many applications in physics and engineering, circular functions appear in combination with other functions or numbers. The following type of relatively simple functions occur frequently in problems involving oscillatory motion. Let $f: R \to R$ be defined by

(6)
$$f(x) = A \sin(\omega x + \epsilon)$$

for all $x \in R$, where A, ω, and ϵ are fixed real numbers, $A > 0, \omega > 0$. To facilitate our investigation of functions of type (6), we shall split the difficulty by first examining three special cases.

Let $f_1: R \to R$ be defined by

(7)
$$f_1(x) = A \sin x$$

for all $x \in R$, $A \in R$, and $A > 0$. According to (7) the value of f_1 at x is $\sin x$ multiplied by A. Now, $\sin(R)$, the image of the sine function, is the interval $[-1, 1]$ and therefore $f_1(R) = [-A, A]$. The graph of f_1 therefore has the appearance of a sine curve (Figure 6.3) with all y-coordinates multiplied by A. The number A is called the *amplitude* of f_1. In Figure 6.15 we exhibit the graphs of three functions of the form (7).

Figure 6.15

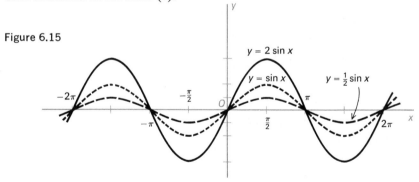

Next let $f_2: R \to R$ be defined by

(8)
$$f_2(x) = \sin \omega x$$

for all $x \in R$, $\omega \in R$, and $\omega > 0$. In this case $f_2(R) = \sin(R)$, but the function f_2

goes through the same values as the sine function ω times "faster" (or "slower" if $\omega < 1$). The function f_2 is periodic with period $2\pi/\omega$. The graph of f_2 has the same appearance as the sine curve except that the oscillations occur with an increased or a decreased frequency according as $\omega > 1$ or $\omega < 1$. Graphs of functions of the type (8) with $\omega = \frac{1}{2}$, 1, and 2 are shown in Figure 6.16.

Figure 6.16

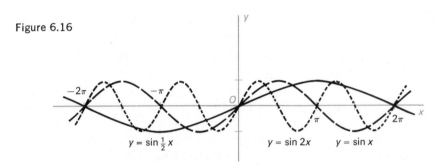

$$y = \sin \tfrac{1}{2}x \qquad y = \sin 2x \qquad y = \sin x$$

Our last special case is $f_3: R \to R$ defined by

(9) $$f_3(x) = \sin(x + \epsilon)$$

where $\epsilon \in R$. For any real number x, the value of f_3 at x is the same as the value of the sine curve at $x + \epsilon$. This means that the graph of f_3 looks like the sine curve moved to the left, if $\epsilon > 0$, or to the right, if $\epsilon < 0$. This is illustrated in the following figure.

Figure 6.17

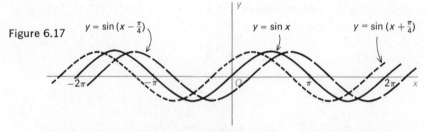

$y = \sin(x - \tfrac{\pi}{4})$ $y = \sin x$ $y = \sin(x + \tfrac{\pi}{4})$

We return now to the function defined in (6): $f(x) = A \sin(\omega x + \epsilon)$. It exhibits simultaneously all the characteristics of the functions f_1, f_2, and f_3. The image $f(R)$ is the interval $[-A, A]$, its period is $2\pi/\omega$, and its graph has the appearance of the sine curve, magnified or diminished vertically and horizontally and displaced to the left or to the right. The number A is called the *amplitude* of f. The number $2\pi/\omega$ is the *period* of f; its reciprocal $\omega/2\pi$ is called the *frequency* of f. Finally, the number $-\epsilon/\omega$ is called the *phase shift* of f. The phase shift represents the horizontal distance the curve $y = A \sin \omega x$ must be moved to coincide with the curve $y = A \sin(\omega x + \epsilon)$. Note in particular that the curve $y = A \sin(\omega x + \epsilon)$ intersects the x-axis at $(-\epsilon/\omega, 0)$.

Example 3.2 (a) Sketch the graph of function $f: R \rightarrow R$ defined by $f(x) = 3 \sin(\pi x - \frac{1}{2}\pi)$. What are the amplitude, the period, and the phase shift of f?

The amplitude of f is 3, its period is $2\pi/\pi = 2$, and its phase shift is $\frac{1}{2}\pi/\pi = \frac{1}{2}$. The graph of f is shown in Figure 6.18.

Figure 6.18

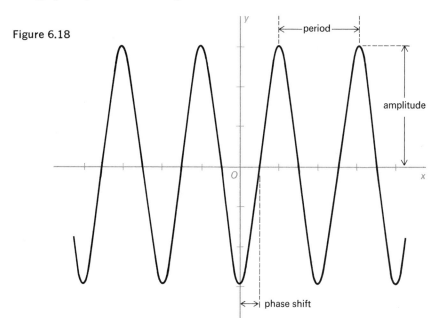

(b) Show that the function defined by $f(x) = -3 \sin(2x + \frac{1}{4}\pi)$ is of the type (6). Determine its amplitude, period, and phase shift.

We have

$$-\sin t = \sin(-t)$$
$$= \sin(\pi - (-t))$$
$$= \sin(t + \pi),$$

and therefore

$$f(x) = -3 \sin(2x + \tfrac{1}{4}\pi)$$
$$= 3 \sin((2x + \tfrac{1}{4}\pi) + \pi)$$
$$= 3 \sin(2x + \tfrac{5}{4}\pi).$$

Hence f has amplitude 3, period $2\pi/2 = \pi$, and phase shift $-\frac{5}{4}\pi/2 = -\frac{5}{8}\pi$.

SYNOPSIS

In this section we defined periodic functions and their periods. We proved that the six circular functions are periodic, and that the period of sine, cosine, cosecant, and secant is 2π, while the period of tangent and cotangent is π. We examined in detail the function $f(x) = A \sin(\omega x + \epsilon)$.

QUIZ

Answer *true* or *false:*

1. The function $g\colon R \to R$ defined by $g(x) = [x]$ is periodic (see Example 3.1(a)).
2. The amplitude of $f(x) = -3 \sin 2x$ is 3.
3. The period of the function f in Question 2 is π.
4. The phase shift of the function f in Question 2 is 0.
5. The frequency of the function f in Question 2 is $\frac{1}{2}$.
6. The function $f(x) = \sin^2 x$ is periodic.
7. The period of the function f in Question 6 is 2π.
8. The period of the function $f(x) = \sin 2\pi x$ is 1.
9. The frequency of the function f in Question 8 is 1.
10. The period of the function $f(x) = \tan 2\pi x$ is 1.

EXERCISES

1. Prove in detail that the cosecant and secant have period 2π.
2. Prove in detail that cotangent has period π.
3. For each of the following functions $f\colon R \to R$, determine the amplitude, period, frequency, and phase shift:
   (a) $4 \sin x$,   (b) $\sin 3x$,
   (c) $\sin \frac{1}{3}x$,   (d) $\sin \pi x$,
   (e) $\sin(x + \frac{1}{2}\pi)$, (f) $\sin(x - 1)$,
   (g) $2 \sin(x + 1)$,   (h) $2 \sin(\pi(x + 1))$,
   (i) $\frac{1}{2}\sin(\pi(x - \frac{1}{2}))$, (j) $3 \sin(2x - 1)$.
4. Sketch the graph of each of the functions in Exercise 3.
5. For each of the following functions find an equivalent formula in the form $f(x) = A \sin(\omega x + \epsilon)$ where $A > 0$ and $\omega > 0$, and determine the phase shift:

   (a) $f(x) = -\sin x$;   (b) $f(x) = -2 \sin\left(x + \frac{\pi}{4}\right)$;

   (c) $f(x) = \sin(-2x)$;   (d) $f(x) = \sin\left(\frac{\pi}{2} - x\right)$;

   (e) $f(x) = -3 \sin(\pi(x + \frac{1}{3}))$;   (f) $f(x) = -\frac{1}{2}\sin(-\frac{1}{2}x - \frac{1}{2}\pi)$.
6. If $f\colon R \to R$ is defined by $f(x) = B \cos(\omega x + \epsilon)$, where $B > 0$ and $\omega > 0$, show that $f(x)$ can be expressed in the form (6). What are the amplitude, period, and phase shift of f?

7. If a and b are real numbers, not both 0, show that there exists a real number β such that $\sin \beta = a/\sqrt{a^2 + b^2}$ and $\cos \beta = b/\sqrt{a^2 + b^2}$.

8. If $f: R \rightarrow R$ is defined by $f(x) = a \cos \omega x + b \sin \omega x$, where $\omega > 0$, a and b are real numbers, not both 0, then show that $f(x)$ can be expressed in the form (6) and determine the amplitude, period, and phase shift of f.
    [*Hint:* Use the result in Exercise 7 and the formula $\sin \lambda \cos \mu + \cos \lambda \sin \mu = \sin (\lambda + \mu)$.]

9. If $f: R \rightarrow R$ is given by the following formula, express $f(x)$ in the form $f(x) = A \sin (\omega x + \epsilon)$, where $A > 0$ and $\omega > 0$:
    (a) $f(x) = \cos 2x$;
    (b) $f(x) = -\cos x$;
    (c) $f(x) = \sin x + \cos x$;
    (d) $f(x) = \sin x - \cos x$;
    (e) $f(x) = 3 \sin 2x + 4 \cos 2x$;
    (f) $f(x) = -3 \cos \left(\pi \left(x - \dfrac{\pi}{4} \right) \right)$;
    (g) $f(x) = 4 \sin 2\pi x - 3 \cos 2\pi x$.

10. If a particle moves along the x-axis so that after t seconds from a fixed initial moment its distance from the origin is

$$x = A \sin (\omega t + \epsilon)$$

units, then the particle is said to describe a *simple harmonic motion*. If $A = 3$ units, $\omega = \pi/2$, and $\epsilon = \pi/4$, compute the distance to the origin when $t = 0$ and at $\frac{1}{2}$-second intervals from $t = \frac{1}{2}$ to $t = 8$.

Complex Numbers

7.1 Complex Numbers

In Chapter 1 we discussed various number systems. Recall that the set Z of integers was not large enough to encompass the solutions of first degree equations with integral coefficients. We remedied this deficiency by embedding Z in the set Q of rational numbers. We saw that even the set Q did not contain the solutions of some simple equations with integral coefficients such as $x^2 - 2 = 0$. We then embedded Q in the set R of all real numbers which contained solutions to many equations of this type. Consider, however, the equation

$$(1) \qquad\qquad x^2 + 1 = 0.$$

It is clear that if x is any real number then $x^2 \geq 0$. Hence, for any real x, the number $x^2 + 1$ is at least 1 and therefore can never be 0. Thus the equation (1) has no solution in the real numbers. As was in the case of the rational numbers, it again becomes expedient to invent a new set of numbers that will contain all the real numbers and also solutions to equations such as (1).

We first introduce a new "number" which we denote by i and which satisfies equation (1):

$$i^2 = -1.$$

We now define a set C of *complex numbers* as the smallest set that contains i and all real numbers, and whose elements satisfy the usual algebraic rules.* In particular, we require that, given any two complex numbers z_1 and z_2, their sum

---

*For a rigorous but more abstract definition of complex numbers that also demonstrates the consistency of our construction in this section, see, e.g., Marcus and Minc, *College Algebra*, Section 1.7.

$z_1 + z_2$ and their product $z_1 z_2$ also belong to C. Thus if b is any real number, the product bi must be in C. If a and b are real the sum of a and bi, $a + bi$, must also be a complex number. Further, if a, b, c, and d are real, then

$$(a + bi) + (c + di) = (a + c) + (b + d)i,$$

and

$$(a + bi)(c + di) = ac + adi + bci + bdi^2$$
$$= (ac - bd) + (ad + bc)i,$$

since $i^2 = -1$ and therefore $bdi^2 = bd(-1) = -bd$. In the above computation we used the "usual algebraic rules": the distributivity, commutativity, and associativity of addition and multiplication (see Section 1.4). We have shown that the sum and the product of complex numbers of the form $a + bi$ is another number of the same form. Note also that any real number a can also be written as $a + 0i$ and that $i = 0 + 1i$. It follows that all complex numbers can be expressed in the form $a + bi$, where both a and b are real. For convenience we now repeat some of the above remarks in the form of a definition.

Definition 1.1 (Complex Numbers) *Complex numbers* are elements of the set

$$(2) \qquad\qquad C = \{z \mid z = a + bi, a, b \in R\},$$

where

$$(3) \qquad\qquad i^2 = -1.$$

Two complex numbers $a + bi$ and $c + di$, where a, b, c, d are real, are equal if and only if $a = c$ and $b = d$. The operations of addition and multiplication of complex numbers are defined as follows:

$$(4) \qquad\qquad (a + bi) + (c + di) = (a + c) + (b + d)i,$$

$$(5) \qquad\qquad (a + bi)(c + di) = (ac - bd) + (ad + bc)i.$$

The number $i = 0 + 1i$ is called the *imaginary unit;* it commutes with every real number. Complex numbers of the form $bi = 0 + bi$, where b is real, are called (pure) *imaginary numbers.*

Example 1.1 Show that for every nonzero complex number z there exists a unique complex number w such that

$$(6) \qquad\qquad zw = wz = 1.$$

Let $z = a + bi$, where a and b are real numbers, not both zero since we assume that $z \neq 0$. Thus $a^2 + b^2 > 0$. Set

(7)
$$w = \frac{a}{a^2 + b^2} + \frac{-b}{a^2 + b^2} i$$

and compute

$$\begin{aligned}
zw &= (a + bi) \times \left(\frac{a}{a^2 + b^2} + \frac{-b}{a^2 + b^2} i \right) \\
&= \left(\frac{aa}{a^2 + b^2} \right) - \left(\frac{b(-b)}{a^2 + b^2} \right) + \left(\frac{a(-b)}{a^2 + b^2} + \frac{ba}{a^2 + b^2} \right) i \\
&= \frac{a^2 + b^2}{a^2 + b^2} + \left(\frac{-ab + ab}{a^2 + b^2} \right) i \\
&= 1 + 0i \\
&= 1.
\end{aligned}$$

We can compute in an analogous way that $wz = 1$, or we can take for granted that $zw = wz$ for any two complex numbers. To see that there exists only one w satisfying (6), suppose that

$$zW = 1$$

and that

$$zw = 1.$$

Then

(8)
$$zw = zW.$$

Since $z \neq 0$, there exists a number w' such that

$$w'z = 1.$$

Multiply both sides of (8) by w' to obtain

$$w'zw = w'zW,$$

i.e.,

$$1w = 1W,$$

or

$$w = W,$$

The unique complex number w is called the *inverse* of z and is written z^{-1} or $\frac{1}{z}$.

Formula (7) tells us that if $z = a + bi \neq 0$, then

(9)
$$z^{-1} = \frac{a}{a^2 + b^2} + \frac{-b}{a^2 + b^2}i.$$

For example, we can compute the inverse of i directly from (9):

$$i^{-1} = (0 + 1i)^{-1}$$
$$= \frac{0}{0^2 + 1^2} + \frac{-1}{0^2 + 1^2}i$$
$$= 0 + (-1)i$$
$$= -i.$$

We can derive (9) in a somewhat more formal way as follows. Suppose it is required to compute z^{-1}, where $z = a + bi \neq 0$. If we write z^{-1} as $1/z$ and then multiply numerator and denominator by the complex number $a - bi$, we have

$$\frac{1}{a + bi} = \frac{a - bi}{(a + bi)(a - bi)}$$
$$= \frac{a - bi}{a^2 + b^2}$$
$$= \frac{a}{a^2 + b^2} + \frac{-b}{a^2 + b^2}i,$$

which is precisely the formula (9).

If z is a complex number and n is a positive integer then

$$z^n = \overbrace{z \cdot z \cdots z}^{n},$$

and

$$z^{-n} = (z^{-1})^n$$
$$= 1/z^n.$$

If $z \neq 0$, we also define

$$z^0 = 1.$$

Example 1.2   (a)   Express $(1 + i)^{-2}$ in the form $a + bi$, where a and b are real.
The expression $(1 + i)^{-2}$ means $(1 + i)^{-1}$ is to be squared. Now we know that

$$(1 + i)^{-1} = \frac{1}{1 + 1} - \frac{i}{1 + 1}$$
$$= \frac{1}{2} - \frac{i}{2}$$
$$= \tfrac{1}{2}(1 - i)$$

$(\text{see } (9))$. Hence

$$
\begin{aligned}
(1 + i)^{-2} &= \left((1 + i)^{-1}\right)^2 \\
&= \left(\tfrac{1}{2}(1 - i)\right)^2 \\
&= \left((\tfrac{1}{2})^2\right)(1 - i)^2 \\
&= \tfrac{1}{4}(1 - 2i + i^2) \\
&= \tfrac{1}{4}(1 - 2i - 1) \\
&= \tfrac{1}{4}(-2i) \\
&= -i/2.
\end{aligned}
$$

More formally, we can compute $(1 + i)^{-2}$ as follows:

$$
\begin{aligned}
(1 + i)^{-2} &= \frac{1}{(1 + i)^2} \\[2mm]
&= \frac{1}{1 + 2i + i^2} \\[2mm]
&= \frac{1}{1 + 2i - 1} \\[2mm]
&= \frac{1}{2i} \\[2mm]
&= \frac{i}{2 \cdot i \cdot i} \\[2mm]
&= \frac{i}{2 \cdot (-1)} \\[2mm]
&= -i/2.
\end{aligned}
$$

(b)   Express $(3 + i)(2 - i)$ in the form $a + bi$, a and b real.
We have

$$
\begin{aligned}
(3 + i)(2 - i) &= 3(2 - i) + i(2 - i) \\
&= 6 - 3i + 2i - i^2 \\
&= 6 - i - (-1) \\
&= 7 - i.
\end{aligned}
$$

(c)   Express $\dfrac{2 + i}{3 - i}$ in the form $a + bi$, a and b real.

We use (9) to compute

$$
\begin{aligned}
\frac{2 + i}{3 - i} &= (2 + i)(3 - i)^{-1} \\[2mm]
&= (2 + i)\left(\frac{3}{3^2 + 1^2} + \frac{1}{3^2 + 1^2}i\right)
\end{aligned}
$$

$$= (2 + i)\left(\frac{3}{10} + \frac{i}{10}\right)$$
$$= \tfrac{6}{10} - \tfrac{1}{10} + (\tfrac{3}{10} + \tfrac{2}{10})i$$
$$= \tfrac{1}{2} + \tfrac{1}{2}i.$$

Definition 1.2 (Complex Conjugate, Absolute Value) If $z = a + bi$ is a complex number, a and b real, then the complex number $a - bi$ is called the *complex conjugate* of z and is denoted by $\bar{z}$, i.e.,

$$\bar{z} = a - bi.$$

The nonnegative square root of the real number $z\bar{z}$ is called the *absolute value* or the *modulus* of z and is denoted by $|z|$:

$$|z| = \sqrt{z\bar{z}}.$$

Observe that

$$z\bar{z} = (a + bi)(a - bi)$$
$$= a^2 - b^2 i^2 + abi - abi$$
$$= a^2 + b^2,$$

and hence

(10) $$|a + bi| = \sqrt{a^2 + b^2}.$$

(We always take the nonnegative square root.)

Notice that if $z = a$, i.e., if z is a real number, then

$$|z| = |a|$$
$$= \sqrt{a^2},$$

which is always a nonnegative number. The absolute value of a real number is often called its *numerical value*. For example, the numerical value of -3 is 3, the numerical value of 0 is 0, etc.

Definition 1.3 (Real and Imaginary Parts) If $z = a + bi \in C$, a and b real, then the real number a is called the *real part* of z and the real number b is called the *imaginary part* of z. This is written

$$\mathrm{Re}(z) = a,$$
$$\mathrm{Im}(z) = b.$$

Example 1.3   Find the real and imaginary parts of

$$z = 4 + (5i)i.$$

If we express z in the form $a + bi$, we see that

$$z = 4 + 5i^2$$
$$= 4 - 5$$
$$= -1.$$

Hence

$$\text{Re}(z) = -1$$

and

$$\text{Im}(z) = 0.$$

In the next theorem we summarize some of the elementary properties of the absolute value and conjugate.

Theorem 1.1   *Let z, z_1, z_2 be complex numbers. Then*

(a) $\text{Re}(z) = \dfrac{z + \bar{z}}{2}$,

(b) $\text{Im}(z) = \dfrac{z - \bar{z}}{2i}$,

(c) $\overline{(z_1 + z_2)} = \bar{z}_1 + \bar{z}_2$,

(d) $\overline{z_1 z_2} = \bar{z}_1 \bar{z}_2$,

(e) $\bar{\bar{z}} = z$,

(f) $|z_1 z_2| = |z_1|\,|z_2|$,

(g) $|z_1 + z_2| \le |z_1| + |z_2|$.

Proof   (a)   Let $z = a + bi$, a and b real, and observe that

$$z + \bar{z} = (a + bi) + (a - bi)$$
$$= 2a.$$

In other words,

$$2\,\text{Re}(z) = z + \bar{z}.$$

(b)   The proof is almost the same as for (a) and is left as an exercise.

(c)   Let $z_1 = a_1 + b_1 i$ and $z_2 = a_2 + b_2 i$. Then

$$\overline{z_1 + z_2} = \overline{(a_1 + a_2) + (b_1 + b_2)i}$$
$$= (a_1 + a_2) - (b_1 + b_2)i$$
$$= (a_1 - b_1 i) + (a_2 - b_2 i)$$
$$= \bar{z}_1 + \bar{z}_2.$$

(d) As in (c),

$$\overline{z_1 z_2} = \overline{(a_1 + b_1 i)(a_2 + b_2 i)}$$
$$= \overline{(a_1 a_2 - b_1 b_2) + (a_1 b_2 + a_2 b_1)i}$$
$$= (a_1 a_2 - b_1 b_2) - (a_1 b_2 + a_2 b_1)i$$
$$= (a_1 - b_1 i)(a_2 - b_2 i)$$
$$= \bar{z}_1 \bar{z}_2.$$

(e) We leave this verification as an exercise.

(f) From Definition 1.2,

$$|z_1 z_2|^2 = (z_1 z_2)(\overline{z_1 z_2})$$
$$= z_1 z_2 \bar{z}_1 \bar{z}_2$$
$$= z_1 \bar{z}_1 z_2 \bar{z}_2$$
$$= |z_1|^2 |z_2|^2.$$

If we take the square root of both sides of this last equation, the result follows.

(g) Using Definition 1.2 again, we have

(11)
$$|z_1 + z_2|^2 = (z_1 + z_2)(\overline{z_1 + z_2})$$
$$= (z_1 + z_2)(\bar{z}_1 + \bar{z}_2)$$
$$= z_1 \bar{z}_1 + z_2 \bar{z}_2 + z_1 \bar{z}_2 + z_2 \bar{z}_1$$
$$= |z_1|^2 + |z_2|^2 + z_1 \bar{z}_2 + \overline{z_1 \bar{z}_2}.$$

The complex number $z_1 \bar{z}_2 + \overline{z_1 \bar{z}_2}$ appearing on the right in (11) is

$$2 \operatorname{Re}(z_1 \bar{z}_2).$$

The absolute value of a complex number is always at least equal to its real part; i.e., if $z = a + bi$, then

$$\operatorname{Re}(z) = a$$
$$\leq \sqrt{a^2}$$
$$\leq \sqrt{a^2 + b^2}$$
$$= |z|.$$

Applying this last remark to the complex number $z_1 \bar{z}_2$, we have

$$2 \operatorname{Re}(z_1 \bar{z}_2) \leq 2|z_1 \bar{z}_2|$$
$$= 2|z_1| |\bar{z}_2|$$
$$= 2|z_1| |z_2|,$$

i.e.,

(12)
$$z_1 \bar{z}_2 + \overline{z_1 \bar{z}_2} \le 2|z_1|\,|z_2|.$$

In the last computation we have used the fact that

$$|z| = |\bar{z}|$$

for any complex number z. Substituting (12) into (11) we have

$$|z_1 + z_2|^2 \le |z_1|^2 + |z_2|^2 + 2|z_1|\,|z_2|$$
$$= (|z_1| + |z_2|)^2.$$

This completes the proof. ∎

Example 1.4   (a)   Find the absolute value (or modulus) of the complex number

$$(1 + i)^2.$$

Using Theorem 1.1 (f), we have

$$|(1 + i)^2| = |(1 + i)|^2$$
$$= (\sqrt{1 + 1})^2$$
$$= 2.$$

(b)   Find the imaginary part of $(2 + i)^3$.
We have
$$(2 + i)^3 = 2^3 + 3 \cdot 2^2 \cdot i + 3 \cdot 2 \cdot i^2 + i^3$$
$$= 8 + 12i - 6 - i$$
$$= 2 + 11i.$$

Hence
$$\mathrm{Im}\big((2 + i)^3\big) = 11.$$

SYNOPSIS

In this section we introduced the set C of complex numbers and developed some elementary rules of computation for complex numbers. In Theorem 1.1 we listed a number of interesting and important facts relating the absolute value, the complex conjugate, and the real and imaginary parts of complex numbers.

QUIZ

Answer *true* or *false:*

In the following questions, x and y are real numbers and z and w are complex numbers.

1. If $x + yi = 0$, then $x = y = 0$.
2. If $z = |z|$, then z must be real.

3. If z is not zero, then $z^{-1} = \dfrac{z}{|z|^2}$.

4. $(1 + i)^{-1} = 1 - i$.
5. $|i| = |-i|$.
6. $|i| = -1$.
7. $\text{Re}(z\bar{z}) \geq 0$.
8. $\text{Im}(z\bar{z}) = 0$.
9. $z = \text{Re}(z) + \text{Im}(z)$.
10. $\text{Re}(zw) = \text{Re}(z)\text{Re}(w)$.

EXERCISES

1. Find the conjugate and the modulus of each of the following complex numbers:

   (a) 2,   (b) -2,
   (c) $2i$,   (d) $-2i$,
   (e) $1 + i$,   (f) $-3 + i$,
   (g) $-3 - i$,   (h) i^2,
   (i) $i + 2$,   (j) $-5i - 7$,
   (k) $\sqrt{3} - \sqrt{3}\,i$,   (l) $-1 + \sqrt{3}\,i$,
   (m) $(3 + i) + (2 - i)$,   (n) $(i - 1) + (2 - i)$,
   (o) $(\sqrt{3} + i) + (2\sqrt{3} - 3i)$,   (p) $-\sqrt{3}\,i/2$.

2. Express the following complex numbers in the form $x + yi$, in which x and y are real:

   (a) i^{-1},   (b) $(2 + i)(3 - i)$,
   (c) $(1 + i)^2$,   (d) $(3 - 2i)(-4 + i)$,

   (e) $(1 + i)^{-1}$,   (f) $\dfrac{2 + 3i}{2 - 3i}$,

   (g) $\dfrac{3 + 2i}{-1 + 5i}$,   (h) $\dfrac{1 - 2i}{-2 + 3i}$,

   (i) $\dfrac{1 + 2i}{2 - i} + \dfrac{2i}{-3 + i}$,   (j) $\dfrac{1 + 2i}{2 - i} \cdot \dfrac{2i}{-3 + i}$,

   (k) $\dfrac{7 - 6i}{1 + i} - \dfrac{3 - i}{2 - 9i}$,   (l) $\dfrac{7 - 6i}{1 + i} \cdot \dfrac{3 - i}{2 - 9i}$,

(m) $3 - 2i + \dfrac{4 - 3i}{2 + i}$,

(n) $\text{Im}((1 + i)^2)$,

(o) $(\text{Re}(i))^2$,

(p) $\dfrac{(2 - i)(1 + 3i)}{(1 + i)(3 + 2i)}$.

3. (a) Show that $\text{Re}(z) = 0$ if and only if $\bar{z} = -z$.

(b) Show that $\text{Im}(z) = 0$ if and only if $z = \bar{z}$.

(c) Show that for any two complex numbers z_1 and z_2, $|z_1 z_2| = 0$ if and only if $z_1 = 0$ or $z_2 = 0$.

(d) Show that $|z^{-1}| = |z|^{-1}$, for any nonzero complex number z.

4. Prove Theorem 1.1(b).

5. Prove Theorem 1.1(e).

6. State the necessary and sufficient conditions for equality to hold in Theorem 1.1(g).

7. Find the inverses of the following complex numbers:

(a) $2i$,

(b) $-3i$,

(c) $1 - i$,

(d) $3 + 4i$,

(e) $-\dfrac{1}{2} + \dfrac{\sqrt{3}}{2}i$,

(f) $\dfrac{1}{\sqrt{2}} + \dfrac{1}{\sqrt{2}}i$,

(g) $\dfrac{\sqrt{3} - i}{1 - \sqrt{3}i}$,

(h) $\left(-\dfrac{1}{2} + \dfrac{\sqrt{3}}{2}i\right)^2$.

8. Show that $\left|\dfrac{z_1}{z_2}\right| = \dfrac{|z_1|}{|z_2|}$.

9. Show that a complex number is 0 if and only if its modulus is 0.

7.2 Trigonometric Form of Complex Numbers

A complex number is completely defined by its real and imaginary parts. In other words, to each complex number $x + yi$ corresponds a unique ordered pair of real numbers (x, y) and to each ordered pair (x, y) of real numbers we can relate a unique complex number $x + yi$. Thus once a pair of mutually perpendicular axes is chosen, we can represent the elements of C as points of the plane (see Figure 7.1).

Figure 7.1

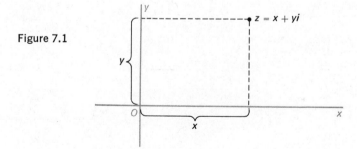

The y-axis in Figure 7.1 is called the *imaginary* axis and the x-axis is called the *real* axis. The reason for this nomenclature is obvious, since all the numbers of the form $z = x$ lie on the x-axis and all the numbers of the form $z = yi$ lie on the y-axis. The representation of complex numbers as indicated in Figure 7.1 is called the *Argand diagram* for C.

Addition of complex numbers has a simple geometric representation in terms of the Argand diagram. Let $z = x_1 + y_1 i$ and $w = x_2 + y_2 i$ be any complex numbers. Then $z + w = (x_1 + x_2) + (y_1 + y_2)i$. Therefore on the Argand diagram the points $P(x_1, y_1)$, $Q(x_2, y_2)$, and $R(x_1 + x_2, y_1 + y_2)$ represent the numbers z, w, and $z + w$. Clearly $OPRQ$ is a parallelogram. Thus if we regard vectors $\overrightarrow{OP}$ and $\overrightarrow{OQ}$ as representing z and w, then $z + w$ is represented by the vector $\overrightarrow{OP} + \overrightarrow{OQ} = \overrightarrow{OR}$ (see Figure 7.2).

Figure 7.2

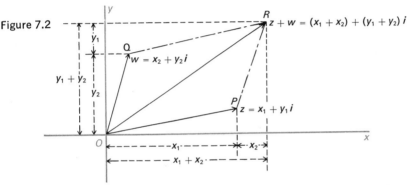

Complex numbers correspond in the Argand diagram to points in the plane that are specified by reference to a pair of mutually perpendicular coordinate axes. Another important representation for complex numbers that has many advantages is obtained by referring the points to a different system of coordinates, the *polar coordinates*.

Suppose that O is a fixed point in a plane and l a fixed line through O called the *polar axis*. A point P may be located by giving its distance r from the point O, together with the angle θ that the line segment OP makes with the fixed line l, measured counterclockwise from l. Thus the point P is specified by the pair (r, θ), the polar coordinates of P (Figure 7.3).

Figure 7.3

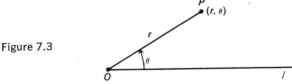

The number r is always taken to be nonnegative and the angle θ may be measured in either radians or degrees. If θ is a negative number, then the angle is measured clockwise starting from l.

If a rectangular coordinate axis system is taken so that the positive x-axis coincides with the line l in Figure 7.3, it is extremely simple to find the relationship between the rectangular and polar coordinates of a point P as in Figure 7.4.

Figure 7.4

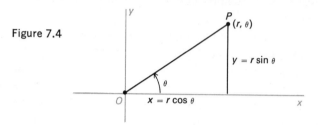

We associate an angle θ with every point P except the origin $(0, 0)$. From the Figure 7.4 we have:

(1)
$$x = r\cos\theta,$$
$$y = r\sin\theta,$$
$$r^2 = x^2 + y^2,$$
$$y/x = \tan\theta, \ x \neq 0.$$

Example 2.1   (a)   Find the polar coordinates of the point whose rectangular coordinates are $(1, -1)$.

From (1) we have

$$1^2 + (-1)^2 = r^2,$$

and hence

$$r = \sqrt{2}.$$

Also from (1),

$$\cos\theta = x/r$$
$$= 1/\sqrt{2},$$

and

$$\sin\theta = y/r$$
$$= -1/\sqrt{2}.$$

Thus $\theta = 315°$. It follows that

$$(\sqrt{2}, 315°)$$

are the polar coordinates of the point $(1, -1)$.

(b)   Show that the point $(r, \theta + k \cdot 2\pi)$ has the same rectangular coordinates for every value of the integer k.

Since the trigonometric functions sin θ and cos θ are periodic of period 2π, it follows from (1) that the rectangular coordinates of the indicated point are

$$x = r \cos (\theta + k \cdot 2\pi)$$
$$= r \cos \theta,$$

for any integer k, and

$$y = r \sin (\theta + k \cdot 2\pi)$$
$$= r \sin \theta,$$

for any integer k.

We return now to the geometrical representation of complex numbers. Let $z = x + yi$ be a complex number represented in the Argand diagram by the point (x, y) (Figure 7.5).

Figure 7.5

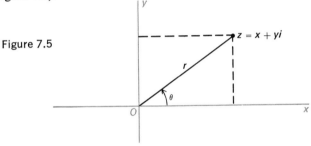

We saw that if we choose the nonnegative x-axis as the polar axis and if the point (x, y) has polar coordinates r and θ, then these coordinates are related as follows:

(2) $$x = r \cos \theta,$$

(3) $$y = r \sin \theta,$$

(4) $$r = \sqrt{x^2 + y^2}.$$

Hence we can write the number $z = x + yi$ in the form

$$z = r \cos \theta + ir \sin \theta$$

or

(5) $$z = r(\cos \theta + i \sin \theta).$$

(Note that in order to avoid ambiguity, we wrote the number i in front of the imaginary part, which is permissible since i commutes with real numbers.)

Formula (5) is called the *trigonometric* or the *polar form* of z. It is clear from (4) that r is equal to $|z|$, the absolute value of z. The angle θ is called the *amplitude* of z and is denoted by amp(z). If $z \neq 0$ then amp(z) is uniquely determined to within a multiple of 360°. This follows immediately from the one–one correspondence between complex numbers and points in a plane. In fact, if $\text{Re}(z) \neq 0$, then

$$\tan\left(\text{amp}(z)\right) = y/x$$
$$= \frac{\text{Im}(z)}{\text{Re}(z)}.$$

Thus if $\text{Re}(z) > 0$, then $\text{amp}(z) = \tan^{-1}\dfrac{\text{Im}(z)}{\text{Re}(z)}$, and if $\text{Re}(z) < 0$, then $\text{amp}(z) = 180° + \tan^{-1}\dfrac{\text{Im}(z)}{\text{Re}(z)}$. If $\text{Re}(z) = 0$, then $\text{amp}(z) = 90°$ if $\text{Im}(z) > 0$, and $\text{amp}(z) = 270°$ if $\text{Im}(z) < 0$. In case $z = 0$, amp(z) is not defined.

Example 2.2   (a) The absolute value of z is 4 and its amplitude is 150°. Find the real and imaginary parts of z.
   The polar form of z is

$$4(\cos 150° + i\sin 150°) = 4\cos 150° + i4\sin 150°$$
$$= -2\sqrt{3} + i2.$$

Hence

$$\text{Re}(z) = -2\sqrt{3},$$

and

$$\text{Im}(z) = 2.$$

(b) Express $z = 5 + i8$ in polar form.
   We compute

$$|z| = \sqrt{5^2 + 8^2}$$
$$= \sqrt{89},$$

and, with the aid of Table 2 in the Appendix,

$$\text{amp}(z) = \tan^{-1}\tfrac{8}{5}$$
$$= 58°.$$

Therefore the polar form of z is

$$\sqrt{89}\,(\cos 58° + i\sin 58°).$$

(c)   Express $z = \dfrac{4 + i12}{1 - i2}$ in polar form.

We first multiply the numerator and the denominator by the complex conjugate of the denominator:

$$z = \frac{(4 + i12)(1 + i2)}{(1 - i2)(1 + i2)}$$

$$= \frac{-20 + i20}{1 + 4}$$

$$= -4 + i4.$$

Hence

$$|z| = \sqrt{(-4)^2 + 4^2}$$

$$= 4\sqrt{2},$$

and

$$amp(z) = 180° + \tan^{-1}(4/-4)$$

$$= 180° + (-45°)$$

$$= 135°.$$

Therefore, the polar form of z is

$$4\sqrt{2}\,(\cos 135° + i \sin 135°).$$

Actually, it is not necessary to memorize any formulas in order to determine whether $amp(z)$ is equal to $\tan^{-1} \dfrac{Im(z)}{Re(z)}$ or $180° + \tan^{-1} \dfrac{Im(z)}{Re(z)}$. In the above example, it suffices to observe that $Im(z) > 0$, $Re(z) < 0$, and therefore z lies in quadrant II, whereas if $Im(z) < 0$ and $Re(z) > 0$, then $\tan^{-1}$ has values in quadrant IV.

The trigonometric representation of complex numbers is an important and powerful device. This will be evident from the following two theorems.

Theorem 2.1   *Let w be the product of nonzero complex numbers $z_1, z_2, \ldots, z_n$. Let the polar form of z_j be $r_j(\cos \theta_j + i \sin \theta_j), j = 1, 2, \ldots, n$. Then*

$$w = s(\cos \Theta + i \sin \Theta),$$

where

$$s = r_1 r_2 \cdots r_n$$

and

$$\Theta = \theta_1 + \theta_2 + \cdots + \theta_n.$$

Proof If $n = 2$, then

$$
\begin{aligned}
w &= z_1 z_2 \\
&= r_1(\cos \theta_1 + i \sin \theta_1) r_2(\cos \theta_2 + i \sin \theta_2) \\
&= r_1 r_2 \big((\cos \theta_1 \cos \theta_2 - \sin \theta_1 \sin \theta_2) + i(\sin \theta_1 \cos \theta_2 + \cos \theta_1 \sin \theta_2)\big) \\
&= r_1 r_2 \big(\cos (\theta_1 + \theta_2) + i \sin (\theta_1 + \theta_2)\big)
\end{aligned}
$$

by formulas (1) and (3), Section 4.1. If $n = 3$ we can use what we have just established for $n = 2$ as follows:

$$
\begin{aligned}
w &= z_1 z_2 z_3 \\
&= (z_1 z_2) z_3 \\
&= [r_1 r_2 \big(\cos (\theta_1 + \theta_2) + i \sin (\theta_1 + \theta_2)\big)] z_3 \\
&= [r_1 r_2 \big(\cos (\theta_1 + \theta_2) + i \sin (\theta_1 + \theta_2)\big)] r_3(\cos \theta_3 + i \sin \theta_3) \\
&= r_1 r_2 r_3 \big((\cos (\theta_1 + \theta_2) \cos \theta_3 - \sin (\theta_1 + \theta_2) \sin \theta_3) \\
&\qquad + i(\sin (\theta_1 + \theta_2) \cos \theta_3 + \cos (\theta_1 + \theta_2) \sin \theta_3)\big).
\end{aligned}
$$

Using formulas (1) and (3) of Section 4.1 with $\alpha = \theta_1 + \theta_2$ and $\beta = \theta_3$ we can continue the above calculation to obtain

$$
w = r_1 r_2 r_3 \big(\cos (\theta_1 + \theta_2 + \theta_3) + i \sin (\theta_1 + \theta_2 + \theta_3)\big).
$$

Clearly we can continue in this fashion repeatedly using formulas (1) and (3) of Section 4.1 to establish the desired result. ∎

Theorem 2.2 (De Moivre's Theorem) *If $z = r(\cos \theta + i \sin \theta)$ and n is a positive integer, then*

$$
z^n = r^n(\cos n\theta + i \sin n\theta).
$$

Proof Set $r_1 = r_2 = \cdots = r_n = r$ and $\theta_1 = \theta_2 = \cdots = \theta_n = \theta$ in Theorem 2.1. Then $s = r^n$, $\Theta = n\theta$, and the result follows. ∎

Example 2.3 (a) Evaluate w^4 if

$$
w = \frac{4 + i12}{1 - i2}.
$$

We found in Example 2.2(c) that

$$
w = 4\sqrt{2}(\cos 135° + i \sin 135°).
$$

Hence, by De Moivre's Theorem,

$$\begin{aligned} w^4 &= (4\sqrt{2})^4\left(\cos\,(4 \times 135°) + i\sin\,(4 \times 135°)\right) \\ &= 1024(\cos 540° + i\sin 540°) \\ &= 1024\left(\cos\,(360° + 180°) + i\sin\,(360° + 180°)\right) \\ &= 1024(-1 + i \cdot 0) \\ &= -1024. \end{aligned}$$

(b)   Express $\sin 3\alpha$ in terms of $\sin \alpha$; express $\cos 3\alpha$ in terms of $\cos \alpha$.
Let $z = \cos \alpha + i\sin \alpha$. Then, by De Moivre's theorem,

(6) $$z^3 = \cos 3\alpha + i\sin 3\alpha.$$

Also, by direct computation,

(7) $$\begin{aligned} z^3 &= (\cos \alpha + i\sin \alpha)^3 \\ &= \cos^3 \alpha - 3\cos \alpha \sin^2 \alpha + i(3\cos^2 \alpha \sin \alpha - \sin^3 \alpha). \end{aligned}$$

Now, by definition (see Definition 1.1, Section 7.1) two complex numbers are equal if and only if their real parts are equal and their imaginary parts are equal. Thus, equating real and imaginary parts in (6) and (7), we have

$$\begin{aligned} \cos 3\alpha &= \cos^3 \alpha - 3\cos \alpha \sin^2 \alpha \\ &= \cos^3 \alpha - 3\cos \alpha(1 - \cos^2 \alpha) \\ &= 4\cos^3 \alpha - 3\cos \alpha, \end{aligned}$$

and

$$\begin{aligned} \sin 3\alpha &= 3\cos^2 \alpha \sin \alpha - \sin^3 \alpha \\ &= 3(1 - \sin^2 \alpha)\sin \alpha - \sin^3 \alpha \\ &= 3\sin \alpha - 4\sin^3 \alpha. \end{aligned}$$

(c)   Let $z_1 = 2i$ and $z_2 = -\sqrt{3} - i$. Evaluate z_1^3 and z_2^3.
Clearly

$$\begin{aligned} (2i)^3 &= 2^3 i^3 \\ &= 8(-i) \\ &= -8i. \end{aligned}$$

We can evaluate z_2^3 in the same way. However, we shall compute it by means of De Moivre's Theorem. First put z_2 in polar form:

$$\begin{aligned} |z_2| &= \sqrt{(-\sqrt{3})^2 + (-1)^2} \\ &= 2, \end{aligned}$$

and

$$\text{amp}(z_2) = 180° + \tan^{-1}\left(\frac{-1}{-\sqrt{3}}\right)$$
$$= 180° + 30°$$
$$= 210°;$$

therefore

$$z_2 = 2(\cos 210° + i \sin 210°).$$

Now, by Theorem 2.2,

$$z_2^3 = 2^3(\cos (3 \times 210°) + i \sin (3 \times 210°))$$
$$= 8(\cos (360° + 270°) + i \sin (360° + 270°))$$
$$= 8(0 + i(-1))$$
$$= -8i.$$

We shall now show how De Moivre's Theorem can be extended to nonpositive exponents. Recall (Definition 4.1, Section 1.4) that if $z \neq 0$, then

$$z^0 = 1,$$

and if n is a positive integer, then

$$z^{-n} = (z^{-1})^n.$$

Theorem 2.3 *If $z = r(\cos \theta + i \sin \theta)$ is in polar form and n is any integer, then*

$$z^n = r^n(\cos n\theta + i \sin n\theta).$$

(In case $n \leq 0$, assume $r \neq 0$.)

Proof For $n > 0$ the theorem is a restatement of Theorem 2.2. If $n = 0$, we have

$$z^0 = 1$$
$$= r^0(\cos 0 \cdot \theta + i \sin 0 \cdot \theta),$$

and thus the theorem holds.
  If $n = -1$, then

$$z^{-1} = \frac{1}{r(\cos \theta + i \sin \theta)}$$
$$= r^{-1} \frac{\cos \theta - i \sin \theta}{(\cos \theta + i \sin \theta)(\cos \theta - i \sin \theta)}$$

(8)
$$= r^{-1} \frac{\cos \theta - i \sin \theta}{\cos^2 \theta + \sin^2 \theta}$$
$$= r^{-1}(\cos \theta - i \sin \theta)$$
$$= r^{-1}(\cos (-1 \cdot \theta) + i \sin (-1 \cdot \theta)),$$

and again the theorem holds.

Lastly, if n is negative, set $n = -m$, $m > 0$. Then, using Theorem 2.2 and formula (8), we have

$$z^n = (z^m)^{-1}$$
$$= (r^m(\cos m\theta + i \sin m\theta))^{-1}$$
$$= r^{-m}(\cos (-m\theta) + i \sin (-m\theta))$$
$$= r^n(\cos n\theta + i \sin n\theta). \; \blacksquare$$

Note that formula (8) can also be obtained directly from formula (9) in Section 7.1:

$$(a + ib)^{-1} = \frac{a}{a^2 + b^2} + i \frac{-b}{a^2 + b^2}.$$

For, if $z = a + ib = r(\cos \theta + i \sin \theta) \neq 0$, then this last formula can be written

$$z^{-1} = \frac{\bar{z}}{|z|^2},$$

and therefore

$$z^{-1} = \frac{r(\cos \theta - i \sin \theta)}{r^2}$$
$$= r^{-1}(\cos (-\theta) + i \sin (-\theta)).$$

We return to Example 2.3(c), in which we exhibited two unequal complex numbers, $2i$ and $-\sqrt{3} - i$, whose cubes are equal. This situation most certainly cannot occur for two distinct real numbers. In our last theorem we show that any nonzero complex number has n distinct nth roots, and we actually obtain a formula which determines all these roots.

Definition 2.1 (Root of a Complex Number) Let n be a positive integer. An nth *root* of a complex number z is a complex number w whose nth power is equal to z. In other words, any number w which satisfies

$$w^n = z,$$

is called an nth root of z.

For example, both $2i$ and $-\sqrt{3} - i$ are cube roots (i.e., 3rd roots) of $-8i$.

Theorem 2.4 *Let* $r(\cos\theta + i\sin\theta)$ *be the polar form of z. If* $z \neq 0$, *then z has exactly n distinct nth roots,* $w_0, \ldots, w_{n-1}$, *given by the formula*

(9) $$w_k = \sqrt[n]{r}\left(\cos\frac{\theta + k \cdot 360°}{n} + i\sin\frac{\theta + k \cdot 360°}{n}\right),$$

$k = 0, 1, \ldots, n - 1$.

Proof   Let us first note that for any integers i and j satisfying $0 \leq i < j \leq n - 1$,

$$\theta \leq \operatorname{amp}(w_i) < \operatorname{amp}(w_j) < \theta + 360°,$$

i.e., the amplitudes of w_i and w_j differ by less than 360°. It follows that $w_0, w_1, \ldots, w_{n-1}$ are all distinct complex numbers. Next, by Theorem 2.2,

$$
\begin{aligned}
w_k^n &= (\sqrt[n]{r})^n\left(\cos\left(n\,\frac{\theta + k \cdot 360°}{n}\right) + i\sin\left(n\,\frac{\theta + k \cdot 360°}{n}\right)\right)\\
&= r\bigl(\cos(\theta + k \cdot 360°) + i\sin(\theta + k \cdot 360°)\bigr)\\
&= r(\cos\theta + i\sin\theta)\\
&= z,
\end{aligned}
$$

for every integer k, $0 \leq k \leq n - 1$. Hence $w_0, w_1, \ldots, w_{n-1}$ are distinct nth roots of z. It can be shown that every nonzero complex number has exactly n distinct nth roots (see Exercise 9 which also gives a more constructive method of arriving at formula (9)). ∎

SYNOPSIS

In this final section we introduced the Argand diagram as a representation for complex numbers. We then introduced polar coordinates and showed that every complex number can be represented in the form $r(\cos\theta + i\sin\theta)$, where r is a nonnegative real number and θ is a (real) angle. We used some of our trigonometric identities to derive De Moivre's Theorem, which gives an important formula for evaluating the polar form of the nth power of a complex number. We then exploited De Moivre's Theorem to prove a formula for the nth roots of a given complex number.

QUIZ

Answer *true* or *false:*
1. $\cos 0° + i\sin 0° = 0$.
2. If n is an integer, then $(\cos\theta - i\sin\theta)^n = \cos(n\theta) - i\sin(n\theta)$.

3. If n is an integer, then $(-\cos \theta + i \sin \theta)^n = -\cos (n\theta) + i \sin (n\theta)$.
4. The absolute value of $-\cos \theta + i \sin \theta$ is 1.
5. The amplitude of $-\cos \theta + i \sin \theta$ is θ.
6. The amplitude of $-\cos \theta + i \sin \theta$ is $-\theta$.
7. If z is a complex number, $z \neq 0$, then $\text{amp}(z\bar{z}) = 0°$.
8. If z is a complex number satisfying $z^n = 1$ for a positive integer n, then $z = 1$.
9. If n is an integer, $n \geq 2$, and z is a nonzero complex number satisfying $z^n = z$, then $z = 1$.
10. If w_1 and w_2 are nth roots of z, then $w_1 w_2$ is an nth root of z^2.

EXERCISES

1. Give a geometric interpretation of Theorem 1.1(g):

$$|z_1 + z_2| \leq |z_1| + |z_2|.$$

2. If z and w are two complex numbers, draw an appropriate Argand diagram exhibiting the complex number $w - z$.
3. Illustrate each of the following by means of an Argand diagram:
   (a) $(1 + i) - (2 + 3i)$;
   (b) $|z| = |w|$ for two complex numbers z and w;
   (c) $\text{Re}(z) = \text{Im}(z)$;   (d) $\text{Re}(z) = 0$;
   (e) $\text{Im}(z) = 0$;   (f) $\text{Re}(z) > 0, \text{Im}(z) > 0$;
   (g) $\text{Re}(z) > 0, \text{Im}(z) < 0$;   (h) $\text{Re}(z) < 0, \text{Im}(z) < 0$;
   (i) $\text{Re}(z) < 0, \text{Im}(z) > 0$.
4. Find the rectangular coordinates of each of the following points which are given in polar coordinates:
   (a) $(4, -90°)$,   (b) $(2, 180°)$,   (c) $(0, \pi/6)$,
   (d) (π, π),   (e) $(1, 30°)$,   (f) $(2, 135°)$,
   (g) $(1, 210°)$,   (h) $(3, 20\pi)$,   (i) $(1, 0)$,
   (j) $(\pi, 60°)$.
5. Express the following complex numbers in polar form (use Table 2 in the Appendix if necessary):
   (a) $1 - i$,   (b) $-i$,
   (c) $\dfrac{7 - i}{1 + i}$,   (d) $-\cos 30° - i \sin 30°$,
   (e) $\cos 35° + i \cos 35°$,   (f) $(\sqrt{3} - i)^8$,
   (g) $\left(\dfrac{(1 - i\sqrt{3})(1 - i)}{2i(\sqrt{3} - 1)} \right)^4$,   (h) $\left(\dfrac{5 - i}{3 + 2i} \div \dfrac{1 - 2i}{2 + i} \right)^5$.
6. (a) Write out a formula for the nth roots of 1.

(b) Find all of the 6th roots of 1; express them in the form $x + yi$, where x and y are real.

(c) Plot the six complex numbers computed in (b) on the Argand diagram.

7. (a) Write out a formula for the nth roots of -1.

(b) Find all the 6th roots of -1.

(c) Plot all 6th roots of -1 on the Argand diagram.

8. (a) Let $z_1 = r_1(\cos \theta_1 + i \sin \theta_1)$ and $z_2 = r_2(\cos \theta_2 + i \sin \theta_2)$ be two complex numbers in polar form, $z_2 \neq 0$. Show directly (without using Theorem 2.3) that

$$z_1/z_2 = (r_1/r_2)\big(\cos (\theta_1 - \theta_2) + i \sin (\theta_1 - \theta_2)\big).$$

(b) Use the above formula to find the polar form of z_1/z_2 if

$$z_1 = -1 + i\sqrt{3} \quad \text{and} \quad z_2 = 1 - i.$$

9. Let n be a positive integer,

$$z = r(\cos \theta + i \sin \theta)$$
$$w = \rho(\cos \varphi + i \sin \varphi),$$

and suppose that $z = w^n$. Use De Moivre's Theorem to find the polar form of w^n. Compare the real and imaginary parts of w^n and z and deduce (9). [*Hint:* $\sin \alpha_1 = \sin \alpha_2$ and $\cos \alpha_1 = \cos \alpha_2$ imply that $\alpha_1 = \alpha_2 + k \cdot 360°$ for some integer k, not necessarily that $\alpha_1 = \alpha_2$.]

10. Find all nth roots of z if:

(a) $n = 3, z = i$;

(b) $n = 6, z = -i$;

(c) $n = 4, z = 3 - 4i$;

(d) $n = 5, z = 2 + i$;

(e) $n = 8, z = \dfrac{7 - i}{1 + i}$.

(In 10(d), (e), use Table 2.)

Answers to Quizzes

and Selected Exercises

CHAPTER 1

QUIZ

1. false
3. true
5. true
7. true

9. false

2. true
4. true
6. false
8. false (Let $X = \{\phi\}$, the set whose only member is the empty set ϕ.)

10. true

EXERCISES

1. ϕ, $\{0\}$, $\{1\}$, $\{2\}$, $\{0, 1\}$, $\{0, 2\}$, $\{1, 2\}$, $\{0, 1, 2\}$ = S.
2. ϕ, $\{b\}$, $\{a\}$, $\{n\}$, $\{a, b\}$, $\{a, n\}$, $\{b, n\}$, $\{a, b, n\}$.
3. Consider the set $X = \{1, \{1\}\}$. Then $\{1\}$ is the required element in X.
5.

$\in$	M	E	T	S	R	ϕ
9	YES	NO	YES	YES	NO	NO
5	YES	NO	NO	NO	YES	NO
0	YES	YES	YES	YES	NO	NO
6	YES	YES	YES	NO	NO	NO
26	YES	YES	NO	NO	YES	NO
-1	NO	NO	NO	NO	NO	NO

243

Section 1.2

QUIZ

1.	true	2.	true	3.	true	4.	true	5.	false
6.	true	7.	true	8.	true	9.	true	10.	true

EXERCISES

1. (a) $X \cup Y = \{1, 3, 4, 5, 7, 9\}$.
   (c) $X \cap Z = \{3, 9\}$.
   (i) $X \cap (Y \cap Z) = \{9\}$.
   (k) $(X \cap Y) \cup Z = \{1, 3, 6, 9\}$.
2. (b) $E \cap M = \{x \mid x = 2m, m \in M\}$.
   (d) $(E \cap T) \cup M = \{x \mid x \in M\} = M$.
7.

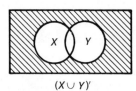 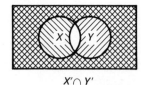

$(X \cup Y)'$ $\qquad\qquad$ $X' \cap Y'$

9. If $n \in Z$, then n is divisible by 6 and hence by both 2 and 3, i.e., $n \in X$ and $n \in Y$. Thus $n \in X \cap Y$. On the other hand, if $n \in X \cap Y$, then n is divisible by 2 and by 3 and hence by $2 \cdot 3 = 6$. Thus $n \in Z$. We have proved that $Z = X \cap Y$.
10. It is not true that $Z = X \cap Y$, because $4 \in X \cap Y$ but $4 \notin Z$.
11. (a) $X = \{1, 2\}$; thus $v(X) = 2$.
14. We know that $v(X) = 83$, $v(Y) = 67$, $v(Z) = 121$, $v(X \cap Y) = 11$, $v(X \cap Z) = 15$, $v(Y \cap Z) = 9$, and $v(X \cap Y \cap Z) = 3$. Thus

$$v(X \cup Y \cup Z) = 83 + 67 + 121 - (11 + 15 + 9) + 3 = 239.$$

Section 1.3

QUIZ

1.	false	2.	false	3.	false	4.	false	5.	true
6.	true	7.	true	8.	false	9.	false	10.	true

EXERCISES

1. (e) The set $Q' \cap R$ is the set of all real numbers which are not rational; i.e., $Q' \cap R$ is the set of all irrational numbers. Thus $N \cup (Q' \cap R)$ is the set consisting of the natural numbers and the irrational numbers.

2. (a), (c), (d).

3. (e) By definition, $1.\overline{414}$ is $1 + .\overline{414} = 1 + .414414414\ldots$. Now $10^3(.\overline{414}) = 414.414414\ldots = 414 + .\overline{414},$ $(10^3 - 1)(.\overline{414}) = 414,$ $.\overline{414} = 414/999$. Thus $1.\overline{414} = 1 + 414/999 = 1413/999 = 157/111$.

   (i) We can easily compute that $.\overline{11} = 1/9$ and $.\overline{22} = 2/9$. Thus $.\overline{11}/.\overline{22} = \frac{1}{2}$.

   (p) 136/567.

5. Suppose that $p/q = \sqrt{3}$, where p and q are integers having no common factors, i.e., p/q is in lowest terms. Then $p^2 = 3q^2$ and hence p^2 is a multiple of 3. It follows that p itself must be a multiple of 3. For, if p is divided by 3, the remainder must be 0, 1, or 2. Suppose that the remainder is 1, i.e., $p = 3m + 1$. Then $p^2 = 9m^2 + 6m + 1 = 3(3m^2 + 2m) + 1 = 3M + 1$. This means p^2 yields a remainder of 1 upon division by 3. Similarly, if $p = 3m + 2$, then $p^2 = 9m^2 + 12m + 4 = 9m^2 + 12m + 3 + 1 = 3(3m^2 + 4m + 1) + 1 = 3M + 1$, and once again, p^2 yields a remainder of 1 upon division by 3. It follows that in fact p must be divisible by 3, say $p = 3m$. Then from $p^2 = 3q^2$, we have $9m^2 = 3q^2$ and hence $q^2 = 3m^2$. Thus by an analogous argument, q must be a multiple of 3. But now p and q have been proved to be multiples of three which contradicts the assertion that p and q have no common factors. Hence $\sqrt{3}$ must be irrational.

6. (b) $.7/.8 = 7/10 \div 8/10 = 7/8 = .875$.

7. (f) $a = 2$ and $b = \frac{1}{2}(\sqrt{2} + \sqrt{8})$, so $b - a = \frac{1}{2}(\sqrt{2} + \sqrt{8}) - 2 = (3/2)\sqrt{2} - 2 = (3\sqrt{2} - 4)/2 > (3(1.4) - 4)/2 = (4.2 - 4)/2 = .2/2 > 0$, which implies from Definition 3.2 that $b > a$.

   (h) $b - a = .\overline{33} - \sqrt{3} + \sqrt{2} > .\overline{33} - 1.74 + 1.41 = .\overline{33} - .33 = .00\overline{33} > 0$. It follows that $b > a$ from Definition 3.2.

Section 1.4

QUIZ

1. true
3. false
5. true
7. false $((2^0)^0 = (1)^0 = 1.)$
9. true

2. true
4. true
6. false (Of what law is it an example?)
8. true
10. false $((-1)^{-(1-1)} = (-1)^{-1} = -1.)$

EXERCISES

1. (a) $2xy + 5$
   (e) $2y^3 - 4xy^2 + yx^2$
   (h) $9x^4 - 16x^2y^2 + 4y^4$
2. (c) a^{-2}
   (e) $2^{-1/4}$
   (g) 2^6
   (i) $-3^{-2/3}$
3. (b) $\dfrac{x^{16/3}}{625y^{16}}$
   (c) $1 - \dfrac{1}{x}$
   (g) $\dfrac{4}{\sqrt[6]{5}}$
4. (c) $\dfrac{2 + \sqrt{2}}{2}$
   (e) $\dfrac{5 + 10\sqrt{2} + 10\sqrt{3} - 2\sqrt{5}}{10}$
5. (d) $\dfrac{4 - \sqrt{6} - 6\sqrt{2} + 3\sqrt{3}}{5}$
   (f) $\dfrac{-\sqrt{x + 1} - \sqrt{2x + 5}}{x + 4}$

CHAPTER 2

Section 2.1

QUIZ

| 1. false | 2. false | 3. true | 4. true | 5. false |
| 6. false | 7. true | 8. false | 9. true | 10. false |

EXERCISES

1. $f(-2) = 3(-2)^2 + 6(-2) - 2 = -2, f(-1) = -5, f(0) = -2, f(1) = 7,$
   $f(2) = 22.$ To see whether f is one–one, assume that $f(x_1) = f(x_2)$, where
   $x_1 \neq x_2$. Then

   $$3x_1^2 + 6x_1 - 2 = 3x_2^2 + 6x_2 - 2,$$
   $$3x_1^2 - 3x_2^2 + 6x_1 - 6x_2 = 0,$$
   $$(x_1^2 - x_2^2) + 2(x_1 - x_2) = 0.$$

Dividing through by $(x_1 - x_2) \neq 0$, we obtain $x_1 + x_2 + 2 = 0$, i.e., $x_1 = -x_2 - 2$. Therefore, for example, $f(3) = f(-5) = 43$, and f is not one–one.

3. (b) $f(X) = R$.

   (d) $f(X) = \{x \mid x \geq 0, x \in R\}$ (i.e., the set of all nonnegative real numbers).

4. (b) $f_1(1) = 1, f_1(2) = 2, f_1(3) = 3; f_2(1) = 1, f_2(2) = 3, f_2(3) = 2;$
      $f_3(1) = 2, f_3(2) = 1, f_3(3) = 3; f_4(1) = 3, f_4(2) = 2, f_4(3) = 1;$
      $f_5(1) = 2, f_5(2) = 3, f_5(3) = 1; f_6(1) = 3, f_6(2) = 1, f_6(3) = 2.$

   (c) $f(1) = 1, f(2) = 1, f(3) = 2.$

5. If $f(x_1) = f(x_2)$, where x_1 and x_2 are different from -1, then $1/(1 + x_1) = 1/(1 + x_2)$ and hence $x_1 + 1 = x_2 + 1, x_1 = x_2$. Since $1/(1 + x)$ is never zero, it follows that f is 1–1. Let y be any real number, not zero. If we set $x = (1 - y)/y$, then it is easy to check that $f((1 - y)/y) = y$. Since $f(-1) = 0$ by definition, it follows that f is onto R.

6. (c) Following the procedure of Example 1.5, $y = f(x_y) = 3x_y - 2$; hence

$$x_y = \frac{1}{3}y + \frac{2}{3} = f^{-1}(y).$$

   (f) Following the same procedure, we have that $y = f(x_y) = x_y^3$; hence $x_y = \sqrt[3]{y} = f^{-1}(y)$.

   (g) Following the same procedure for each of the cases separately, we find

$$f^{-1}(y) = \begin{cases} y - 2 & \text{if } y \leq 3, \\ \frac{1}{2}y - \frac{1}{2} & \text{if } y \geq 3. \end{cases}$$

Section 2.2

QUIZ

1. true	2. false	3. false	4. true	5. false
6. true	7. true	8. true	9. false	10. true

EXERCISES

2. (d)

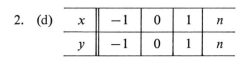

x	-1	0	1	n
y	-1	0	1	n

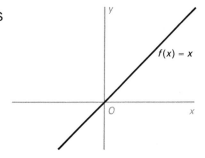

(h)

x	0	2/3
y	2	0

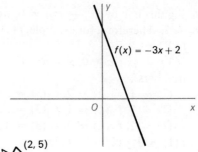

3.

x	0	2	7
y	1	5	0

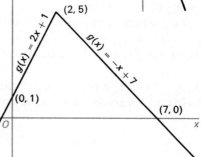

4.

x	0	1	2	3	4
y	2	1	0	2	4

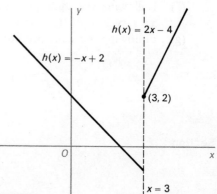

5. (a)

x	-2	-1	0	1	2
y	2	$\frac{1}{2}$	0	$\frac{1}{2}$	2

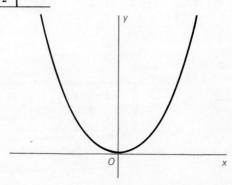

(c)

x	−4	−2	−1	−$\frac{1}{2}$	−$\frac{1}{4}$	0	$\frac{1}{4}$	$\frac{1}{2}$	1	2	4
y	−$\frac{1}{4}$	−$\frac{1}{2}$	−1	−2	−4	0	4	2	1	$\frac{1}{2}$	$\frac{1}{4}$

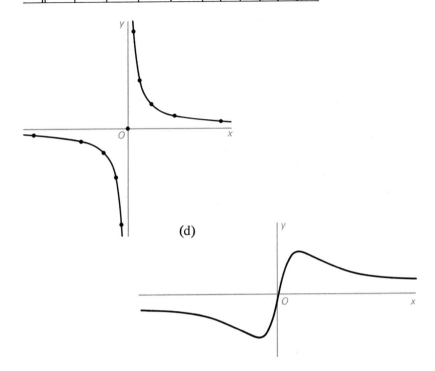

(d)

6. (e) Following Example 2.2, we find $m = (3 - (-2))/(2 - 1) = 5$, $b = (-2) - 1 \cdot 5 = -7$. Hence the function is $f(x) = 5x - 7$.

(h) $m = \dfrac{(-2) - (-3)}{1 - (-2)} = \dfrac{1}{3}$,

$b = (-3) - (-2) \cdot \dfrac{1}{3} = -3 + \dfrac{2}{3} = \dfrac{-7}{3}$,

$f(x) = \dfrac{1}{3}x - \dfrac{7}{3}$.

(n) $m = \dfrac{1/2 - 0}{0 - (-3)} = \dfrac{1/2}{3} = \dfrac{1}{6}$,

$b = 0 - (-3) \cdot \dfrac{1}{6} = \dfrac{1}{2}$,

$f(x) = \dfrac{1}{6}x + \dfrac{1}{2}$.

Section 2.3

QUIZ

1. true
3. false
5. true
7. true
9. false (It is 0.)

2. false $\left(\text{It is } \sqrt{(2-1)^2 + (2-1)^2} = \sqrt{2}.\right)$
4. false (The slope is $-8/5$.)
6. true
8. true
10. true

EXERCISES

1. (b) Using Theorem 3.1, we have

$$\gamma = \sqrt{(3 - (-1))^2 + (-4 - 0)^2}$$
$$= \sqrt{16 + 16}$$
$$= 4\sqrt{2}.$$

(g)
$$\gamma = \sqrt{(3 - (-1))^2 + (-4 - (-2))^2}$$
$$= 2\sqrt{5}.$$

2. (b) In Example 3.3(b) we were given the equation of the midpoint of the line segment joining any two points. In this case we get

$$\left(\frac{1+3}{2}, \frac{1+(-4)}{2}\right) = \left(2, -\frac{3}{2}\right).$$

(d) This time we have the point

$$\left(\frac{3+(-1)}{2}, \frac{1+4}{2}\right) = \left(1, \frac{5}{2}\right).$$

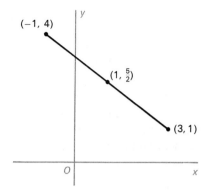

3. (a) The x-intercept is $-b/m = -5/3$, the y-intercept is $b = 5$, and the slope is $m = 3$.

    (c) Rewriting the equation as

    $$y = -\frac{3}{2}x + 3,$$

    we immediately find that the x-intercept is 2, the y-intercept is 3, and the slope is $-3/2$.

    (g) Rewriting the equation we get

    $$y = -\frac{5}{9}x + \frac{11}{9},$$

    which has x-intercept $11/5$, y-intercept $11/9$, and slope $-5/9$.

4. (b) Following the procedure of Example 3.2, we compute

    $$m = (0 - 0)/(1 - 0) = 0,$$

    so the equation of the line is

    $$y = b.$$

    Since $(0, 0)$ lies on the line, $b = 0$ and the equation of the line becomes

    $$y = 0,$$

    i.e., the line is the x-axis.

    (e) We compute that $m = (-1 - 5)/(-1 - 3) = 3/2$, so the equation is of the form

    $$y = \frac{3}{2}x + b.$$

    Since the point $(-1, -1)$ lies on the line, we easily find that $b = \frac{1}{2}$.

(h) Again we compute that $m = (6 - 1)/(4 - 2) = 5/2$, which gives an equation of the form

$$y = \frac{5}{2}x + b.$$

It is now easy to see that $b = -4$.

5. (b) From Theorem 3.3 we know that two lines are perpendicular if and only if the product of their slopes is equal to -1. The slope of the line $y = -x$ is -1; hence the required line will have an equation of the form

$$y = x + b,$$

i.e., $m = 1$. Since the line passes through $(0, 0)$, both the x- and y-intercepts are 0, so the line will have the equation

$$y = x.$$

(d) Rewriting the equation of the given line as

$$y = -\frac{2}{3}x - \frac{5}{3},$$

we know that any line perpendicular to it must be of the form

$$y = \frac{3}{2}x + b.$$

But as in part (b), we get $b = 0$.

(g) We rewrite the equation as

$$y = -\frac{3}{2}x + 3;$$

thus the required equation is

$$y = \frac{2}{3}x.$$

6. (b) Let us attempt to find a positive constant γ such that $\gamma x - \gamma y = \gamma$, where $\gamma^2 + (-\gamma)^2 = 1$. We can solve for γ:

$$2\gamma^2 = 1,$$

$$\gamma^2 = \frac{1}{2},$$

$$\gamma = \frac{1}{\sqrt{2}},$$

since γ must be positive. Hence the normal form of the equation is

$$\frac{1}{\sqrt{2}}x - \frac{1}{\sqrt{2}}y = \frac{1}{\sqrt{2}}.$$

(h) Following the procedure in part (b), we first rewrite the equation as

$$y - 5x = 2.$$

Again we want to find a positive constant γ such that $\gamma y - 5\gamma x = 2\gamma$, where $\gamma^2 + (-5\gamma)^2 = 1$, i.e.,

$$\gamma^2 + 25\gamma^2 = 1,$$
$$26\gamma^2 = 1,$$
$$\gamma = \frac{1}{\sqrt{26}}.$$

Hence the normal form of the equation is

$$\frac{1}{\sqrt{26}}y - \frac{5}{\sqrt{26}}x = \frac{2}{\sqrt{26}}.$$

7. We can rewrite the equation of the line l as

$$y = -\frac{\alpha}{\beta}x + \frac{d}{\beta}.$$

The slope of any line l_1 perpendicular to l must be β/α, and so the equation of l_1 will be of the form

$$y = \frac{\beta}{\alpha}x + b,$$

where b is some constant. We can rewrite this equation as

$$\beta x - \alpha y = c,$$

where c is now the constant $-b\alpha$. If l_1 goes through the origin we already know from Exercise 5 that the y-intercept is $b = 0$, so in this case,

$$c = -b\alpha = 0.$$

8. (c) We can put this equation in normal form following the procedure in Exercise 6: If $\gamma < 0$ is such that $3\gamma x - 4\gamma y = -\gamma$ where $(3\gamma)^2 + (-4\gamma)^2 = 1$, then $9\gamma^2 + 16\gamma^2 = 1$, and

$$\gamma = -\frac{1}{5}.$$

Hence the equation becomes

$$-\frac{3}{5}x + \frac{4}{5}y = \frac{1}{5},$$

so the required line has the equation

$$\frac{4}{5}x + \frac{3}{5}y = 0,$$

or

$$4x + 3y = 0.$$

(f) We rewrite this equation in normal form to obtain $2\gamma x + 2\gamma y = 8\gamma$ where $(2\gamma)^2 + (2\gamma)^2 = 1, \gamma > 0$. Hence

$$4\gamma^2 + 4\gamma^2 = 1,$$
$$\gamma = \frac{1}{2\sqrt{2}},$$

so the equation becomes

$$\frac{1}{\sqrt{2}}x + \frac{1}{\sqrt{2}}y = \frac{4}{\sqrt{2}}.$$

The perpendicular line going through the origin therefore has the equation

$$\frac{1}{\sqrt{2}}x - \frac{1}{\sqrt{2}}y = 0$$

or

$$x - y = 0.$$

(j) The equation of this line in normal form is

$$-\frac{4}{5}x - \frac{3}{5}y = \frac{2}{5}.$$

Thus our required line has the equation

$$-\frac{3}{5}x + \frac{4}{5}y = 0$$

or

$$3x - 4y = 0.$$

9. (a) We can rewrite this equation as

$$y = -\frac{2}{3}x + \frac{5}{3}.$$

Thus the required line has an equation of the form

$$y = \frac{3}{2}x + b.$$

Since the line must pass through $(0, 0)$, we know $b = 0$, and the equation becomes

$$2y = 3x.$$

(d) It is easy to see that this equation is already in normal form. Hence we know that the perpendicular line passing through $(0, 0)$ will have the equation $x = 0$.

(f) We easily find that the required line has the form

$$y = \frac{3}{2}x + b,$$

and since it passes through the point $(3, 2)$ the equation becomes

$$y = \frac{3}{2}x - \frac{5}{2}.$$

11. (a) We put the equation of the line into normal form to obtain $\gamma x + 2\gamma y = \gamma$, where $\gamma^2 + (2\gamma)^2 = 1, \gamma > 0$. Thus

$$\gamma^2 + 4\gamma^2 = 1,$$

$$\gamma = \frac{1}{\sqrt{5}}.$$

Hence the equation of the line is

$$\frac{1}{\sqrt{5}}x + \frac{2}{\sqrt{5}}y = \frac{1}{\sqrt{5}}.$$

We now compute the required distance:

$$\alpha x_0 + \beta y_0 - d = -\left(\left(\frac{1}{\sqrt{5}}\right)(1) + \left(\frac{2}{\sqrt{5}}\right)(-1) - \frac{1}{\sqrt{5}}\right)$$

$$= \frac{2}{\sqrt{5}}.$$

(c)   The normal form of the line is $2\gamma y - \gamma x = \gamma$, where $(2\gamma)^2 + (-\gamma)^2 = 1$, $\gamma > 0$, i.e.,

$$4\gamma^2 + \gamma^2 = 1,$$

$$\gamma = \frac{1}{\sqrt{5}}.$$

Hence the equation is

$$\frac{2}{\sqrt{5}} y - \frac{1}{\sqrt{5}} x = \frac{1}{\sqrt{5}}.$$

The distance between the point and the line is therefore

$$\alpha x_0 + \beta y_0 - d = -\left(\left(\frac{-1}{\sqrt{5}}\right)(4) + \left(\frac{2}{\sqrt{5}}\right)(-3) - \frac{1}{\sqrt{5}}\right)$$

$$= -\left(\frac{-(4 + 6 + 1)}{\sqrt{5}}\right)$$

$$= \frac{11}{\sqrt{5}}.$$

(e)   The equation written in normal form is $-\gamma x + 3\gamma y = -2\gamma$, where $(-\gamma)^2 + (3\gamma)^2 = 1$, $\gamma < 0$, i.e., $\gamma = -1/\sqrt{10}$. Hence the equation is

$$\frac{1}{\sqrt{10}} x - \frac{3}{\sqrt{10}} y = \frac{2}{\sqrt{10}}.$$

We can now easily find the required distance:

$$\alpha x_0 + \beta y_0 - d = \left(\frac{1}{\sqrt{10}}\right)(-1) + \left(\frac{-3}{\sqrt{10}}\right)(-3) - \frac{2}{\sqrt{10}}$$

$$= \frac{6}{\sqrt{10}}.$$

(h) Putting the equation into normal form, we get

$$-\frac{2}{\sqrt{13}}x - \frac{3}{\sqrt{13}}y = \frac{1}{\sqrt{13}},$$

so the distance is

$$\alpha x_0 + \beta y_0 - d = \left(\frac{-2}{\sqrt{13}}\right)(-1) + \left(\frac{-3}{\sqrt{13}}\right)(-2) - \frac{1}{\sqrt{13}}$$

$$= \frac{7}{\sqrt{13}}.$$

12. (b)

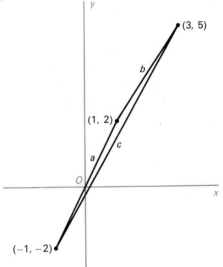

The midpoint of side a is

$$\left(\frac{-1+1}{2}, \frac{-2+2}{2}\right) = (0, 0),$$

and using the distance formula, the length of the median to side a is

$$\sqrt{(3-0)^2 + (5-0)^2} = \sqrt{9 + 25}$$

$$= \sqrt{34}.$$

Similarly, the midpoint of side b is

$$\left(\frac{3+1}{2}, \frac{5+2}{2}\right) = \left(2, \frac{7}{2}\right),$$

and the length of the median to b is

$$\sqrt{(2 - (-1))^2 + \left(\frac{7}{2} - (-2)\right)^2} = \sqrt{9 + \frac{121}{4}} = \frac{\sqrt{157}}{2}.$$

The midpoint of c is

$$\left(\frac{3 + (-1)}{2}, \frac{5 + (-2)}{2}\right) = \left(1, \frac{3}{2}\right),$$

so the length of the median to c is

$$\sqrt{(1 - 1)^2 + \left(2 - \frac{3}{2}\right)^2} = \frac{1}{2}.$$

13. (b)   Following the procedure in Exercise 4, we find the equations of the lines containing the medians to a, b, and c, respectively:

$$y = \frac{5}{3}x,$$

$$y = \frac{11}{6}x - \frac{1}{6},$$

$$x = 1.$$

The point of intersection of $x = 1$ and $y = \frac{11}{6}x - \frac{1}{6}$ is $(1, \frac{5}{3})$, which obviously satisfies the equation $y = \frac{5}{3}x$. Hence the three medians meet at the single point $(1, \frac{5}{3})$.

17.   Using the distance formula, it is easy to see that the two distances are equal, for the distance between (a, u) and $\left(\dfrac{a + c}{2}, \dfrac{u + v}{2}\right)$ is

$$\sqrt{\left(a - \frac{a + c}{2}\right)^2 + \left(u - \frac{u + v}{2}\right)^2} = \sqrt{\left(\frac{a - c}{2}\right)^2 + \left(\frac{u - v}{2}\right)^2},$$

and the distance between $\left(\dfrac{a + c}{2}, \dfrac{u + v}{2}\right)$ and (c, v) is

$$\sqrt{\left(\frac{a + c}{2} - c\right)^2 + \left(\frac{u + v}{2} - v\right)^2} = \sqrt{\left(\frac{a - c}{2}\right)^2 + \left(\frac{u - v}{2}\right)^2}.$$

18. (a) The y-axis has the equation $x = 0$; hence any line parallel to it will have an equation of the form $x = b$. In this case we have $x = 4$.

    (c) A line equally distant from two points will be the perpendicular bisector of the segment joining them. The line passing through $(0, 1)$ and $(1, 0)$ has equation $y = -x + 1$. The midpoint of the line segment joining $(0, 1)$ and $(1, 0)$ is $((1 + 0)/2, (0 + 1)/2) = (1/2, 1/2)$. Therefore, it is easy to see that the equation of the required line is $y = x$.

    (e) The slope of the line is $m = (4 - (-1))/(2 - (-1)) = 5/3$, so the equation of the line is of the form

$$y = \frac{5}{3}x + b.$$

Using the point $(-1, -1)$, we find $b = y - (5/3)x = -1 + 5/3 = 2/3$; hence the desired equation is

$$3y = 5x + 2.$$

    (g) Following the same procedure as in part (c), we find that the equation of the line joining the two given points is

$$y = -\frac{9}{5}x - \frac{2}{5},$$

while the midpoint of the line segment joining them is

$$\left(\frac{-3 + 2}{2}, \frac{5 + (-4)}{2}\right) = \left(-\frac{1}{2}, \frac{1}{2}\right).$$

The perpendicular bisector will have slope $\frac{5}{9}$ and pass through the point $(-\frac{1}{2}, \frac{1}{2})$; hence its equation is

$$9y = 5x + 7.$$

Section 2.4

QUIZ

1. true	2. false	3. true	4. false	5. true
6. false	7. true	8. true	9. true	10. true

EXERCISES

1. (a) 3   (c) 1   (e) $\frac{1}{2}$

   (g) $\frac{1}{2}$ (i) 5   (k) 2

   (m) $\frac{1}{2}$ (o) 1   (q) 4

   (s) $\frac{4}{9}$ (u) 100

2. (a) $4^{3/2} = (4^{1/2})^3 = 2^3 = 8.$

   (c) $\left(\dfrac{1}{9}\right)^{1/2} = \dfrac{1^{1/2}}{9^{1/2}} = \dfrac{1}{3}.$

   (e) $1/8$ (g) 16   (i) $1/5$

   (k) 4   (m) 1000   (o) 8

   (q) .216

3. (a) $x = 3.$ (c) $x = -1/3.$

   (e) $x = -1.$ (g) $x = 3/4.$

   (i) $x = 4/3.$ (k) $x = 0.$

   (m) $x = 1.$

4. (a) $x = 1/2.$ (c) $x = 3/2.$

   (e) $x = 1/4.$ (g) $x = 5/2.$

   (h) $0.04 = \dfrac{4}{100} = \dfrac{1}{25} = \dfrac{1}{(-5)^2} = (-5)^{-2}$; hence $x = -2.$

   (k) $x = -2.$

   (m) $x = -1/3.$

6. (c) As in Example 4.1(c), we apply formula (1):

$$x^{1/2} = 5,$$
$$x = 25.$$

   (g) Following the same procedure as in (c), we find $x = 16.$

   (j) We have $x = \log_3 (1/27) = \log_3 1 - \log_3 27 = 0 - 3 = -3.$

   (o) As in Example 4.1(b), we have $x = 2^{-3} = 1/8.$

   (r) $x = 1/8.$

7. (j) antilog $(\log 2 + \log 3) = $ antilog $(\log (2 \cdot 3))$

                             $= $ antilog $(\log 6)$

                             $= 6.$

8. (a) $\log 6 = \log (3 \cdot 2)$

             $= \log 3 + \log 2$

             $= .7781.$

(c) log 18 = log (3 · 6)
$$= \log 3 + \log 6$$
$$= .4771 + .7781$$
$$= 1.2552.$$

(e) $\log \dfrac{8}{9} = \log 8 - \log 9$
$$= \log (2^3) - \log (3^2)$$
$$= 3 \log 2 - 2 \log 3$$
$$= .9030 - .9542$$
$$= \bar{1}.9488.$$

(g) We know from (d) that log 1.5 = .1761; hence log .015 = $\bar{2}.1761$. Therefore,
$$\log .045 = \log (3 \times .015)$$
$$= \log 3 + \log .015$$
$$= .4771 + \bar{2}.1761$$
$$= \bar{2}.6532.$$

(i) It is easy to see that since
$$\log 8 = \log 2^3$$
$$= 3 \log 2$$
$$= .9030,$$

then
$$\log 0.8 = \bar{1}.9030.$$

Hence
$$\log (1.6) = \log (2 \times 0.8)$$
$$= \log 2 + \log 0.8$$
$$= .3010 + \bar{1}.9030$$
$$= .2040.$$

9. (c) $\log \sqrt{51}$
(e) $\log \sqrt[3]{(1/8)} = \log (1/2).$
(h) $\log (30/25) = \log (6/5).$

Section 2.5

QUIZ

1. false	2. false	3. true	4. false	5. true
6. false	7. true	8. true	9. true	10. true

EXERCISES

1. (d) 4.0899   (g) $\overline{4}.4771$ (r) .7789

2. (d) .708

3. (j) .1097   (r) 13,800

4. (a) 10.11   (c) .07603   (e) 5236
   (g) 14.03   (i) 5.996   (k) 23.54

5. (a) 54.28   (c) .006100   (e) .000002259
   (g) .3310   (k) .000001842   (n) .01873
   (p) 2.249   (r) 2.019   (t) .7483
   (v) .3825

6. (a) 1080   (c) .5359   (e) 670.3
   (g) 15.18   (i) .08804   (k) .7012

7. (a) Given $8^x = 31$ we can write

$$\log 8^x = \log 31,$$
$$x \log 8 = \log 31.$$

Hence

$$x = \frac{\log 31}{\log 8}$$
$$= \frac{1.4914}{.9031}$$
$$= 1.651.$$

(d) We can rewrite this equation as

$$2x \log 5 = \log 15,$$
$$2x = \frac{\log 15}{\log 5},$$
$$x = \frac{1}{2}\left(\frac{\log 15}{\log 5}\right)$$
$$= \frac{1}{2}\left(\frac{1.1761}{.6990}\right)$$
$$= \frac{1}{2}(1.682)$$
$$= .841.$$

(h)   We rewrite the equation to obtain

$$5.72^x = (1.73)(51) + 62,$$

$$x = \frac{\log (1.73 \times 51 + 62)}{\log 5.72}$$

$$= \frac{\log (88.23 + 62)}{\log 5.72}$$

$$= \frac{\log 150.23}{\log 5.72}$$

$$= \frac{2.1767}{.7574}.$$

Here we can use logarithms to compute the approximate value of x.

Numbers	Logs	Antilogs
2.176	.3379	
÷ .7574	− $\overline{1}$.8794	
	.4585	$x = 2.874$

CHAPTER 3

Section 3.1

QUIZ

1. false $(0 \le \sin \alpha \le 1.)$ 2.   false
3. false   (see Table 2.)   4.   false
5. false   6.   false
7. true   8.   true
9. false   10.   true

EXERCISES

3.   (a)   Using the Pythagorean Theorem, we see that

$$a^2 = 5^2 - 3^2 = 16,$$
$$a = 4.$$

We further know that $\sin A = a/c = 4/5$; hence $A = 53°$ and $B = 90° - A = 37°$.

(c) $c = \sqrt{2}$, $A = B = 45°$.

(e) $b = \sqrt{15}$, $A = 14°30'$, $B = 75°30'$.

(g) It is obvious that $B = 90° - A = 50°$. We know that

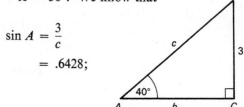

$$\sin A = \frac{3}{c}$$

$$= .6428;$$

hence $c = 4.67$. We can now compute the value of b:

$$b^2 = c^2 - 3^2$$
$$= 21.81 - 9$$
$$= 12.81;$$

therefore $b = 3.58$.

(i) $B = 40°30'$, $b = 2.8$, $c = 4.3$.

(k) $A = 70°$, $a = 2.09$, $c = 2.22$.

5. We bisect the 80° angle to get a right-angled triangle with hypotenuse 50 inches in length. We denote half the length of the chord by c, and compute c:

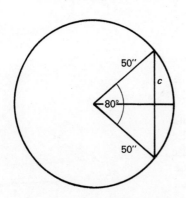

$$\sin 40° = \frac{c}{50},$$

$$c = 50 \sin 40°$$

$$= 32.14 \text{ in.}$$

Thus the length of the chord is $2c = 64.28$ in.

8. (a) We know from elementary geometry that the area of a triangle is equal to one half the base times the height. As in Problem 5, we compute half the length of the chord subtending a 72° angle:

$$\sin 36° = \frac{c}{1},$$

$$c = .5878 \text{ in.}$$

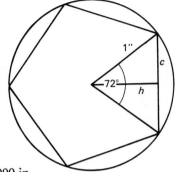

We can also compute h:

$$\sin 54° = \frac{h}{1},$$

$$h = .8090 \text{ in.}$$

Hence the area of the isosceles triangle is

$$\frac{1}{2}(2c)h = ch$$
$$= (.5878)(.8090)$$
$$= .476 \text{ square inches.}$$

Therefore the area of the pentagon is $5(.476) = 2.380$ square inches.

11. It is easy to see that

$$173 = d - d'$$
$$= \frac{h}{\tan 30°} - \frac{h}{\tan 45°}$$
$$= h\left(\frac{1}{\tan 30°} - \frac{1}{\tan 45°}\right)$$
$$= h\left(\frac{1}{.5774} - \frac{1}{1}\right)$$
$$= h\left(\frac{.4226}{.5774}\right)$$
$$= h(.732);$$

thus $h = \dfrac{173}{.732} = 236$ ft.

Section 3.2

QUIZ

1. false	2. false	3. true	4. true	5. false
6. true	7. true	8. true	9. false	10. true

EXERCISES

1. (e) $-.1745$ radians

2. (b)

$$2\pi \text{ radians} = 360°;$$

thus

$$-2\pi \text{ radians} = -360°.$$

(h)

$$1 \text{ radian} = \frac{180}{\pi} \text{ degrees};$$

thus

$$3 \text{ radians} = \frac{3 \cdot 180}{\pi} \text{ degrees}$$

$$= \frac{540}{\pi} \text{ degrees}.$$

(j)

$$1 \text{ radian} = \frac{180}{\pi} \text{ degrees};$$

thus

$$\pi^2 \text{ radians} = \pi^2 \cdot \frac{180}{\pi} \text{ degrees}$$

$$= 180\pi \text{ degrees}.$$

3. (b) $\sin\left(\dfrac{-3\pi}{2}\right) = 1$, $\cos\left(\dfrac{-3\pi}{2}\right) = 0$, $\tan\left(\dfrac{-3\pi}{2}\right)$ is undefined.

(j) We know that

$$\sin(\alpha + k \cdot 360°) = \sin\alpha,$$

or

$$\sin(\alpha + k \cdot 2\pi) = \sin\alpha.$$

Let $k = -1$; then $\sin(\alpha - 2\pi) = \sin\alpha$, so

$$\sin\left(\frac{13\pi}{6} - 2\pi\right) = \sin\frac{\pi}{6}$$

$$= \sin\frac{13\pi}{6}.$$

We know that $\sin \dfrac{\pi}{6} = \dfrac{1}{2}$; thus $\sin \dfrac{13\pi}{6} = \dfrac{1}{2}$. Likewise,

$$\cos \frac{13\pi}{6} = \cos \frac{\pi}{6} = \frac{\sqrt{3}}{2},$$

and

$$\tan \frac{13\pi}{6} = \frac{\sin \dfrac{13\pi}{6}}{\cos \dfrac{13\pi}{6}}$$

$$= \frac{\frac{1}{2}}{\frac{1}{2}\sqrt{3}}$$

$$= \frac{1}{\sqrt{3}}.$$

4. (g)
$$\sec 58° = \frac{1}{\cos 58°}$$

$$= \frac{1}{\sin (90° - 58°)}$$

$$= \frac{1}{\sin 32°}$$

$$= \frac{1}{0.53}$$

$$= 1.88.$$

(i)
$$\tan 148° = -\tan (180° - 148°)$$

$$= -\tan 32°$$

$$= -\frac{\sin 32°}{\cos 32°}$$

$$= -\frac{0.53}{0.85}$$

$$= -0.62.$$

(k)
$$\sin 122° = \sin (180° - 122°)$$

$$= \sin 58°$$

$$= 0.85.$$

5. From Quiz Questions 6 and 7 we have

$$\sin (\alpha + k \cdot 180°) = -\sin \alpha,$$
$$\cos (\alpha + k \cdot 180°) = -\cos \alpha.$$

Thus

$$\tan(\alpha + k \cdot 180°) = \frac{\sin(\alpha + k \cdot 180°)}{\cos(\alpha + k \cdot 180°)}$$

$$= \frac{-\sin\alpha}{-\cos\alpha}$$

$$= \frac{\sin\alpha}{\cos\alpha}$$

$$= \tan\alpha.$$

6. (a) We know from the table (10) of the preceding section that $\sin 30° = \frac{1}{2}$. From formula (10) we also have

$$\sin(180° - 30°) = \sin 150° = \sin 30° = \tfrac{1}{2}.$$

Finally, from (3) we have

$$x = 30° + k \cdot 360° \quad \text{or} \quad x = 150° + k \cdot 360°,$$

where k is an integer.

(e) In Example 2.2 we showed that $\cos\beta = 0$ if and only if $\beta = 90° + k \cdot 180°$, for some integer k. Therefore we have

$$2x = 90° + k \cdot 180°,$$
$$x = 45° + k \cdot 90°.$$

(h) From table (10), Section 3.1, we have $\tan 60° = \sqrt{3}$ and therefore by the identity in Exercise 5,

$$\tan(60° + k \cdot 180°) = \tan 60° = \sqrt{3}$$

for any integer k. Hence

$$30° - x = 60° + k \cdot 180°,$$

i.e.,

$$x = -30° - k \cdot 180°,$$

where k is an integer. Of course, x can be expressed equivalently in the form

$$x = 150° + h \cdot 180°$$

where h is any integer.

8. We know that

$$\alpha + \beta = 180° - \gamma;$$

thus

$$\cos (\alpha + \beta) = \cos (180° - \gamma)$$
$$= -\cos \gamma$$

by formula (11).

Section 3.3

QUIZ

1. false	2. false	3. true	4. false	5. true
6. false	7. false	8. true	9. false	10. true

EXERCISES

(In Exercises 2, 3, and 4 we compute the answer to three-figure accuracy.)

1. (b) $\cos (-325°) = \cos (360° + (-325°))$
$$= \cos 35°$$
$$= .8192.$$

(h) $\cot 200° = \cot (180° - (-20°))$
$$= -\cot (-20°)$$
$$= \cot 20°$$
$$= \tan (90° - 20°)$$
$$= \tan 70°$$
$$= 2.748.$$

(k) $\sin 117°30' = \sin (180° - 117°30')$
$$= \sin 62°30'$$
$$= .8870.$$

(s) $\tan 181°30' = \tan (181°30' - 180°)$
$$= \tan 1°30'$$
$$= .0262.$$

2. (a) $S = 5.4.$
(c) $S = 11700.$
(e) $S = 0.000287.$

3. (a) $M = 38.5$
(c) $M = -0.000191.$
(e) $M = -2.43.$

4. (a) 179,
(c) 13.9,
(e) 130,
(g) 0.00101,
(i) 37.8.

Because the computation of (k) is slightly more difficult, we will illustrate our methods. Let us first compute 4.82 cos 98° using logs. Now, cos 98° = cos (180° − 82°) = −cos 82° is a negative number, so

$$4.82 \cos 98° = -(4.82 \cos 82°).$$

However, we can compute

$$\begin{aligned} \log (4.82 \cos 82°) &= \log 4.82 + \log \cos 82° \\ &= .6830 + \bar{1}.1436 \\ &= \bar{1}.8266. \end{aligned}$$

Further,

$$\begin{aligned} \sqrt[3]{4.82 \cos 98°} &= \sqrt[3]{-4.82 \cos 82°} \\ &= -\sqrt[3]{4.82 \cos 82°}, \end{aligned}$$

and

$$\begin{aligned} \log \sqrt[3]{4.82 \cos 82°} &= \frac{1}{3} \log (4.82 \cos 82°) \\ &= \frac{1}{3} (\bar{1}.8266) \\ &= \bar{1}.9422; \end{aligned}$$

so $\sqrt[3]{4.82 \cos 98°} = -0.88$. Therefore the required number is

$$\sqrt{5.63 - 0.88} = \sqrt{4.75}$$

and again using logs,

$$\begin{aligned} \log \sqrt{4.75} &= \frac{1}{2} \log 4.75 \\ &= \frac{1}{2} (.6767) \\ &= .3384, \end{aligned}$$

or $\sqrt{4.75} = 2.18$.
(m) 0.0204,
(o) 2.21,
(r) −3.95,
(t) 0.517,
(v) 34.5,
(x) 60.3,
(z) 883,000.

5. (b) $\log 46.92 = \log 46.90 + (\log 47.00 - \log 46.90)\dfrac{46.92 - 46.90}{47.00 - 46.90}$

$= 1.6712 + (.0009)\dfrac{2}{10}$

$= 1.6712 + .0002$

$= 1.6714;$

$\log (\sin 25°15')$

$= \log (\sin 25°) + \big(\log (\sin 25°30') - \log (\sin 25°)\big)\dfrac{25°15' - 25°}{25°30' - 25°}$

$= \bar{1}.6260 + (\bar{1}.6340 - \bar{1}.6260)\dfrac{15}{30}$

$= \bar{1}.6260 + (.0080)\tfrac{1}{2}$

$= \bar{1}.6260 + .0040$

$= \bar{1}.6300.$

Numbers	Logs	Antilogs
46.92		1.6714
$\times \sin^3 25°15'$	$3 \times \bar{1}.6300$	$+\ \bar{2}.8900$
		.5614
		3.643

$\text{Antilog} (.5614) = 3.64 + (3.65 - 3.64)\dfrac{.5614 - .5611}{.5623 - .5611}$

$= 3.64 + (.01)\dfrac{3}{12}$

$= 3.64 + .003$

$= 3.643.$

(e) 5.159

Section 3.4

QUIZ

1. false 2. false 3. true 4. true 5. false
6. true 7. false 8. true 9. false 10. true

EXERCISES

1. (f) 45°
    (h) 180°
    (r) 135°

3. (e) 130°30′
   (i) 101°
4. (g) We know that

$$\sin \frac{2\pi}{3} = \frac{\sqrt{3}}{2};$$

therefore

$$\sin^{-1} \sin \frac{2\pi}{3} = \frac{\pi}{3},$$

since $\frac{\pi}{3} \in \left[-\frac{\pi}{2}, \frac{\pi}{2} \right]$.

5. (d) We want to find an angle t such that

$$\cos t = \frac{2}{3},$$

and thus

$$\sin \left(\cos^{-1} \frac{2}{3} \right) = \sin t.$$

We can easily see from a properly constructed right triangle (using the Pythagorean Theorem) that $\sin t = \dfrac{\sqrt{5}}{3}$.

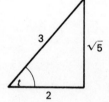

(f) Since $\cos^{-1} \frac{1}{2} = 60°$, we have

$$\cot (\cos^{-1} \tfrac{1}{2}) = \cot 60°$$
$$= \tan 30°$$
$$= \frac{1}{\sqrt{3}}.$$

(j) Let $\sin^{-1} \frac{1}{3} = \alpha$. Thus α is the angle in the right-angled triangle with hypotenuse 3 and opposite side 1.

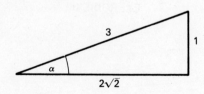

Using the Pythagorean theorem we find that the adjacent side is

$$\sqrt{3^2 - 1^2} = 2\sqrt{2}.$$

Hence

$$\tan\left(\frac{\pi}{2} - \sin^{-1}\frac{1}{3}\right) = \tan\left(\frac{\pi}{2} - \alpha\right)$$
$$= \cot\alpha$$
$$= 2\sqrt{2}.$$

6. (e) We know there is an angle t such that $\sin t = x$.

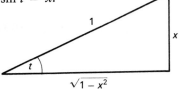

Hence we easily find that

$$\tan(\sin^{-1}x) = \tan t$$
$$= \frac{x}{\sqrt{1 - x^2}}.$$

(h) Let

$$\tan^{-1}\sqrt{x^2 - 1} = y.$$

Then

$$\sec^2 y = 1 + \tan^2 y$$
$$= 1 + \left(\tan(\tan^{-1}\sqrt{x^2 - 1})\right)^2$$
$$= 1 + (\sqrt{x^2 - 1})^2$$
$$= x^2.$$

Therefore

$$\sec(\tan^{-1}\sqrt{x^2 - 1}) = \sec y$$
$$= x.$$

7. (a) In this case, we know that either $\sin x = 0$ or $\cos x = 0$; hence $x = \frac{1}{2}n\pi$, where n is an integer.

(c) We can substitute $1 - \sin^2 x$ for $\cos^2 x$ to obtain

$$3\sin x = 2(1 - \sin^2 x)$$

or

$$2\sin^2 x + 3\sin x - 2 = 0.$$

Using the quadratic equation we get

$$\sin x = \frac{1}{2} \quad \text{or} \quad \sin x = -2.$$

Since -2 is not in the range of the sine function, the only possibility is

$$\sin x = \frac{1}{2},$$

so

$$x = 30° + k \cdot 2\pi \quad \text{or} \quad x = 150° + k \cdot 2\pi,$$

where k is an integer.

CHAPTER 4

Section 4.1

QUIZ

1. true	2. true	3. false	4. true	5. false
6. true	7. true	8. false	9. false	10. false

EXERCISES

6. (c) By formula (18) we have

$$\cos 58° + \cos 12° = 2 \cos \frac{58° + 12°}{2} \cos \frac{58° - 12°}{2}$$
$$= 2 \cos 35° \cos 23°.$$

(i) $\cos 66° + \sin 57° = \sin 24° + \sin 57°$
$$= 2 \sin 40°30' \cos 16°30'.$$

8. (d) $\cos 36° \cos 40° = \cos 40° \cos 36°$
$$= \tfrac{1}{2} \left(\cos (40° + 36°) + \cos (40° - 36°) \right)$$
$$= \tfrac{1}{2} \left(\cos 76° + \cos 4° \right).$$

(i) $2 (\cos 3x \cos 2x) = 2[\tfrac{1}{2}(\cos (3x + 2x) + \cos (3x - 2x))]$
$$= \cos 5x + \cos x.$$

9. (b) Substitute $1 - 2 \sin^2 x$ in place of $\cos 2x$ in the left-hand side and obtain the quadratic equation

$$6 \sin^2 x + (1 - 2 \sin^2 x) = 2,$$
$$4 \sin^2 x = 1,$$
$$\sin^2 x = \tfrac{1}{4},$$
$$\sin x = \pm\tfrac{1}{2}.$$

Hence, in the interval $[0°, 360°)$, x can equal $30°$, $150°$, $210°$, or $330°$.

(g) Apply formula (16) to the left-hand side:

$$\sin 2x + \sin 4x = 2 \sin 3x \cos x.$$

Thus the equation becomes

$$2 \sin 3x \cos x = \cos x,$$

i.e.,

$$(2 \sin 3x - 1) \cos x = 0.$$

Hence one of the factors on the left-hand side must be zero:

$$2 \sin 3x - 1 = 0$$

or

$$\cos x = 0.$$

If $2 \sin 3x - 1 = 0$, then $\sin 3x = \tfrac{1}{2}$ and $\big($see Exercise 6(a), Section 3.2$\big)$

$$3x = 30° + k \cdot 360° \quad \text{or} \quad 3x = 150° + k \cdot 360°,$$

where k is an integer. In the interval $[0°, 360°)$, x can therefore take the values

$$10°, 130°, 250°, \quad \text{or} \quad 50°, 170°, 290°.$$

If $\cos x = 0$, then $x = 90°$ or $270°$.

10. (b) $\sin^4 \alpha = \tfrac{3}{8} - \tfrac{1}{2} \cos 2\alpha + \tfrac{1}{8} \cos 4\alpha,$

11. Using Theorem 1.1 and table (10) of Section 3.1, we find

$$\cos 15° = \cos (45° - 30°)$$
$$= \cos 45° \cos 30° + \sin 45° \sin 30°$$
$$= \frac{1}{\sqrt{2}} \cdot \frac{\sqrt{3}}{2} + \frac{1}{\sqrt{2}} \cdot \frac{1}{2}$$
$$= \frac{\sqrt{3} + 1}{2\sqrt{2}}.$$

Section 4.2

QUIZ

1. true	2. false	3. true	4. true	5. true
6. false	7. true	8. false	9. false	10. true

EXERCISES

1. (d) This is an identity:

$$
\begin{aligned}
\cos 2\alpha &= \sin (90° - 2\alpha) \\
&= \sin \big(2(45° - \alpha)\big) \\
&= 2 \sin (45° - \alpha) \cos (45° - \alpha) \\
&= 2 \sin (45° - \alpha) \cos (\alpha - 45°).
\end{aligned}
$$

Alternatively, the right-hand side can be transformed by means of the formula in Exercise 7(a), Section 4.1:

$$
\begin{aligned}
2 \sin (45° - \alpha) \cos (\alpha - 45°) &= \sin \big((45° - \alpha) + (\alpha - 45°)\big) \\
&\quad + \sin \big((45° - \alpha) - (\alpha - 45°)\big) \\
&= \sin 0° + \sin (90° - 2\alpha) \\
&= \cos 2\alpha.
\end{aligned}
$$

2. (d)
$$
\begin{aligned}
\frac{1 + \tan \alpha}{1 - \tan \alpha} &= \frac{1 + \dfrac{\sin \alpha}{\cos \alpha}}{1 - \dfrac{\sin \alpha}{\cos \alpha}} \\[2ex]
&= \frac{\dfrac{\cos \alpha + \sin \alpha}{\cos \alpha}}{\dfrac{\cos \alpha - \sin \alpha}{\cos \alpha}} \\[2ex]
&= \frac{\cos \alpha + \sin \alpha}{\cos \alpha - \sin \alpha}.
\end{aligned}
$$

(f)
$$
\begin{aligned}
\sin^4 \alpha - \cos^4 \alpha &= (\sin^2 \alpha + \cos^2 \alpha)(\sin^2 \alpha - \cos^2 \alpha) \\
&= 1 \cdot (\sin^2 \alpha - \cos^2 \alpha) \\
&= (\sin^2 \alpha - (1 - \sin^2 \alpha)) \\
&= 2 \sin^2 \alpha - 1.
\end{aligned}
$$

(h) $\sin (45° + \alpha) \sec (45° - \alpha) = \dfrac{\sin (45° + \alpha)}{\cos (45° - \alpha)}$

$$= \dfrac{\cos (90° - (45° + \alpha))}{\cos (45° - \alpha)}$$

$$= \dfrac{\cos (45° - \alpha)}{\cos (45° - \alpha)}$$

$$= 1$$

$$= \dfrac{\cos (45° + \alpha)}{\cos (45° + \alpha)}$$

$$= \dfrac{\cos (45° + \alpha)}{\sin (90° - (45° + \alpha))}$$

$$= \dfrac{\cos (45° + \alpha)}{\sin (45° - \alpha)}$$

$$= \cos (45° + \alpha) \operatorname{cosec} (45° - \alpha).$$

(j) $\dfrac{1 - \cos \alpha}{1 + \cos \alpha} = \left(\dfrac{1 - \cos \alpha}{1 + \cos \alpha}\right)\left(\dfrac{1 - \cos \alpha}{1 - \cos \alpha}\right)$

$$= \dfrac{(1 - \cos \alpha)^2}{1 - \cos^2 \alpha}$$

$$= \dfrac{1 - 2 \cos \alpha + \cos^2 \alpha}{\sin^2 \alpha}$$

$$= \dfrac{1}{\sin^2 \alpha} - \dfrac{2 \cos \alpha}{\sin^2 \alpha} + \dfrac{\cos^2 \alpha}{\sin^2 \alpha}$$

$$= \operatorname{cosec}^2 \alpha - 2 \cot \alpha \operatorname{cosec} \alpha + \cot^2 \alpha$$

$$= (\operatorname{cosec} \alpha - \cot \alpha)^2.$$

3. (d) $\dfrac{\tan \dfrac{\alpha + \beta}{2}}{\tan \dfrac{\alpha - \beta}{2}} = \dfrac{\left(\sin \dfrac{\alpha + \beta}{2}\right) \Big/ \left(\cos \dfrac{\alpha + \beta}{2}\right)}{\left(\sin \dfrac{\alpha - \beta}{2}\right) \Big/ \left(\cos \dfrac{\alpha - \beta}{2}\right)}$

$$= \dfrac{\sin \dfrac{\alpha + \beta}{2} \cos \dfrac{\alpha - \beta}{2}}{\cos \dfrac{\alpha + \beta}{2} \sin \dfrac{\alpha - \beta}{2}}$$

$$= \dfrac{2 \sin \dfrac{\alpha + \beta}{2} \cos \dfrac{\alpha - \beta}{2}}{2 \cos \dfrac{\alpha + \beta}{2} \sin \dfrac{\alpha - \beta}{2}}$$

$$= \dfrac{\sin \alpha + \sin \beta}{\sin \alpha - \sin \beta}.$$

(g)
$$\frac{\cos 2\beta - \sin 2\beta}{\sin \beta \cos \beta} = \frac{\cos^2 \beta - \sin^2 \beta - 2 \sin \beta \cos \beta}{\sin \beta \cos \beta}$$

$$= \frac{\cos \beta}{\sin \beta} - \frac{\sin \beta}{\cos \beta} - \frac{2 \sin \beta \cos \beta}{\sin \beta \cos \beta}$$

$$= \cot \beta - \tan \beta - 2.$$

(m)
$$\frac{\sin 10\alpha - \sin 4\alpha}{\sin 4\alpha + \sin 2\alpha} = \frac{2 \cos \dfrac{10\alpha + 4\alpha}{2} \sin \dfrac{10\alpha - 4\alpha}{2}}{2 \sin \dfrac{4\alpha + 2\alpha}{2} \cos \dfrac{4\alpha - 2\alpha}{2}}$$

$$= \frac{2 \cos 7\alpha \sin 3\alpha}{2 \sin 3\alpha \cos \alpha}$$

$$= \frac{\cos 7\alpha}{\cos \alpha}.$$

(p)
$$\text{RHS} = \sin \theta + \sin 2\theta$$

$$= \frac{(2 - 2 \cos \theta)(\sin \theta + \sin 2\theta)}{(2 - 2 \cos \theta)}$$

$$= \frac{2 \sin \theta + 2 \sin 2\theta - 2 \sin \theta \cos \theta - 2 \sin 2\theta \cos \theta}{2 - 2 \cos \theta}$$

$$= \frac{2 \sin \theta - 2 \sin 2\theta \cos \theta + \sin 2\theta}{2 - 2 \cos \theta}$$

$$= \frac{2 \sin \theta - (\sin 3\theta + \sin \theta) + \sin 2\theta}{2 - 2 \cos \theta}$$

$$= \frac{\sin \theta - \sin 3\theta + \sin 2\theta}{2 - 2 \cos \theta} = \text{LHS.}$$

CHAPTER 5

Section 5.1

QUIZ

1. true	2. false	3. true	4. true	5. true
6. true	7. false	8. true	9. true	10. false

EXERCISES

1. (a) $B = 58°, b = 11.2, c = 13.2.$
   (e) $A = 50°, B = 40°, a = 50.4.$
   (i) $A = 23°30', b = 7.50, c = 8.17.$
   (k) $A = 22°30', B = 67°30', b = 31.4, c = 34.$
   (m) $B = 51°, a = 6.05, b = 7.47, c = 9.61.$
   (o) $A = 47°, B = 43°, a = 4.96, c = 6.78.$
   (s) $B = 58°30', a = 4.17, b = 6.81, c = 7.98.$

2. (b) In the isosceles triangle AOB, $\angle AOB = \dfrac{360°}{5} = 72°$ and $\angle OAB = \angle OBA = 54°$. Let OC be the altitude of triangle AOB.

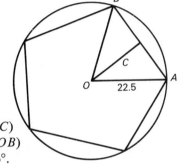

Then

$$AB = 2AC$$
$$= 2OA \sin (\angle AOC)$$
$$= 2OA \sin \tfrac{1}{2}(\angle AOB)$$
$$= 2 \times 22.5 \sin 36°.$$

Numbers	Logs	Antilogs
2	0.3010	
$\times$ 22.5	+ 1.3522	
$\times \sin 36°$	+ $\bar{1}$.7692	
	1.4224	26.5

Therefore the perimeter is equal to $5 \times 26.5 = 132.5.$

3. (c) Since the perimeter is 1.25, the length of a side of the decagon is $\dfrac{1.25}{10} = 0.125.$

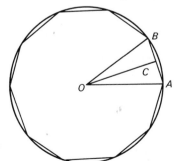

Also $\angle AOB = \dfrac{360°}{10} = 36°$. Let OC be the altitude of the isosceles triangle AOB. Then

$$OA = \dfrac{AC}{\sin(\angle AOC)}$$

$$= \dfrac{\frac{1}{2}AB}{\sin\frac{1}{2}(\angle AOB)}$$

$$= \dfrac{.125}{2\sin\dfrac{36°}{2}}.$$

Numbers	Logs	Antilogs
0.125		$\bar{1}.0969$
$\div \begin{cases} 2 \\ \times \sin 18° \end{cases}$ $\begin{aligned} & 0.3010 \\ &+ \bar{1}.4900 \\ \hline & \bar{1}.7910 \end{aligned}$	$\begin{aligned} &- \bar{1}.7910 \\ \hline & \bar{1}.3059 \end{aligned}$	$OA = .202$

7. In the figure, A and B represent the two positions of the driver, and C and D the two stationary vehicles. Then $CD = 300$ yds. In the right-angled triangle BAD:

$$AD = AB \tan 27°.$$

In the right-angled triangle BAC:

$$AC = AB \tan 17°.$$

Thus

$$\dfrac{AD}{AC} = \dfrac{\tan 27°}{\tan 17°}.$$

Now $AD = AC + 300$, and therefore

$$\dfrac{AC + 300}{AC} = \dfrac{\tan 27°}{\tan 17°},$$

i.e.,

$$\dfrac{300}{AC} = \dfrac{\tan 27°}{\tan 17°} - 1.$$

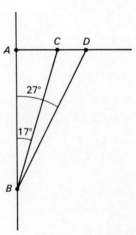

Numbers	Logs	Antilogs
tan 27°	$\bar{1}.7072$	
÷ tan 17°	$-\bar{1}.4853$	
	$\overline{0.2219}$	1.667

Hence

$$\frac{300}{AC} = 0.667,$$

and

$$AC = 450.$$

Therefore

$$AD = 750,$$

and in triangle BAD

$$AB = AD \tan (\angle BDA)$$
$$= 750 \tan 63°$$
$$= 1472.$$

Thus the speed of the car was 1472 yards per minute, or approximately 50.2 m.p.h.

Section 5.2

QUIZ

1. false   2. true   3. false   4. true   5. true
6. true   7. false   8. false   9. false   10. true

EXERCISES

1. (b) We know that
$$C = (180° - (A + B))$$
$$= (180° - 82°)$$
$$= 98°.$$

Using the law of sines, we have

$$\frac{1}{\sin 38°} = \frac{b}{\sin 44°} = \frac{c}{\sin 98°}.$$

We compute that $b = \dfrac{\sin 44°}{\sin 38°}$.

Numbers	Logs	Antilogs
$\sin 44°$	$\bar{1}.8418$	
$\div \sin 38°$	$- \bar{1}.7893$	
	$.0525$	1.13

Hence $b = 1.13$. Also, $c = \dfrac{\sin 98°}{\sin 38°}$.

Numbers	Logs	Antilogs
$\sin 98°$	$\bar{1}.9958$	
$\div \sin 38°$	$- \bar{1}.7893$	
	$.2065$	1.61

Thus $c = 1.61$.

(c) We know immediately that

$$A = 180° - (B + C)$$
$$= 180° - (37° + 100°)$$
$$= 43°.$$

Using the law of sines,

$$\frac{a}{\sin A} = \frac{b}{\sin B},$$

or

$$b = \frac{a \sin B}{\sin A}$$
$$= \frac{17.6 \sin 37°}{\sin 43°}.$$

Numbers	Logs	Antilogs
17.6	1.2455	
$\times \sin 37°$	$+ \bar{1}.7795$	
	1.0250	
$\div \sin 43°$	$- \bar{1}.8338$	
	1.1912	15.5

Hence $b = 15.5$, and we can compute c using the same procedure:

$$\frac{a}{\sin A} = \frac{c}{\sin C},$$

or

$$c = \frac{a \sin C}{\sin A}$$

$$= \frac{17.6 \sin 100°}{\sin 43°}$$

$$= \frac{17.6 \sin 80°}{\sin 43°}.$$

Numbers	Logs	Antilogs
17.6	1.2455	
× sin 80°	+ $\overline{1}$.9934	
	1.2389	
÷ sin 43°	− $\overline{1}$.8338	
	1.4051	25.4

Hence $c = 25.4$.

(e) We first compute the third angle

$$B = 180° - (31°30' + 90°30') = 58°,$$

and use the law of sines:

$$\frac{.034}{\sin 58°} = \frac{a}{\sin 31°30'} = \frac{c}{\sin 90°30'}.$$

Therefore

$$a = \frac{.034 \sin 31°30'}{\sin 58°},$$

and

$$c = \frac{.034 \sin 90°30'}{\sin 58°}.$$

Numbers	Logs	Antilogs
$\begin{cases} .034 \\ × \sin 31°30' \end{cases}$	$\overline{2}$.5315 + $\overline{1}$.7181	
	$\overline{2}$.2496	
÷ sin 58°	− $\overline{1}$.9284	
	$\overline{2}$.3212	$a = .021$

Numbers	Logs	Antilogs
$\Big\{$ .034 $\times \sin 90°30'$ $(= \sin 89°30')$	$\overline{2}.5315$ $+$ .0000 $\overline{2}.5315$	
$\div \sin 58°$	$-$ $\overline{1}.9284$ $\overline{2}.6031$	$c = .040$

(g)   From the law of cosines, we have

$$\cos A = \frac{b^2 + c^2 - a^2}{2bc}$$

$$= \frac{7^2 + 10^2 - 5^2}{2(7)(10)}$$

$$= \frac{49 + 100 - 25}{140}$$

$$= \frac{124}{140}$$

$$= .8857,$$

$$A = 27°30';$$

$$\cos B = \frac{a^2 + c^2 - b^2}{2ac}$$

$$= \frac{25 + 100 - 49}{100}$$

$$= \frac{76}{100}$$

$$= .7600,$$

$$B = 40°30';$$

and

$$C = 180° - (A + B)$$
$$= 180° - (27°30' + 40°30')$$
$$= 180° - 68°$$
$$= 112°.$$

2. (c)   Using the law of sines

$$\frac{a}{\sin A} = \frac{b}{\sin B},$$

$$\sin A = \frac{a \sin B}{b}$$

$$= \frac{5 \sin 70°}{7}.$$

Numbers	Logs	Antilogs
5	.6990	
× sin 70°	+ $\overline{1}$.9730	
	.6720	
÷ 7	− .8451	
	$\overline{1}$.8269	.671

Hence $\sin A = .671$, and therefore $A = 42°$ or $A = 138°$. But A cannot be 138° since $A + B + C = 180°$ and thus $A = 180° - B - C < 180° - B = 180° - 70° = 110°$.
Therefore,

$$C = 180° - (A + B)$$
$$= 180° - 112°$$
$$= 68°.$$

Again, by the law of sines

$$\frac{c}{\sin C} = \frac{b}{\sin B},$$

$$c = \frac{b \sin C}{\sin B}$$

$$= \frac{7 \sin 68°}{\sin 70°}.$$

Numbers	Logs	Antilogs
7	.8451	
× sin 68°	+ $\overline{1}$.9672	
	.8123	
÷ sin 70°	− $\overline{1}$.9730	
	.8393	6.91

Hence, $c = 6.91$.

(g)   Using the law of sines we have

$$\frac{94.9}{\sin 110°} = \frac{97.2}{\sin B},$$

i.e.,

$$\sin B = \frac{97.2 \sin 110°}{94.9}.$$

Numbers	Logs	Antilogs
$\begin{cases} 97.2 \\ \times \sin 110° \\ (= \sin 70°) \end{cases}$	1.9877 $+ \overline{1}.9730$	
	1.9607	
$\div 94.9$	$- 1.9773$	
	$\overline{1}.9834$	$.963 = \sin B$

We find from Table 2 that $B = 70°30'$ or $B = 105°30'$. But

$$B = 180° - A - C < 180° - A = 180° - 110° = 70°.$$

This implies that there are no solutions, i.e., that there exists no triangle ABC with $A = 110°$, $b = 97.2$ and $a = 94.9$.

6. Since 40 minutes is equivalent to 2/3 of an hour, the jet plane has traveled $600 \times (2/3) = 400$ miles and the propeller plane has traveled $270 \times (2/3) = 180$ miles by the time they meet. We can now draw a diagram of the routes of the planes.

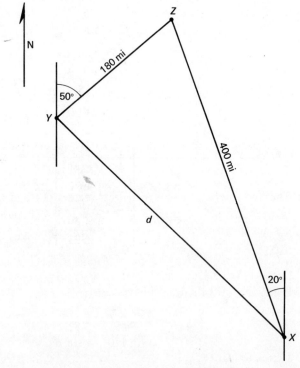

It can easily be found that the angle at Z (the angle at which the planes meet) is 70°. The distance d between the two towns X and Y can be calculated directly from the law of cosines:

$$d^2 = 180^2 + 400^2 - 2(180)(400) \cos 70°$$
$$= 160{,}000 + 32{,}400 - 144{,}000 \cos 70°$$
$$= 192{,}400 - 49{,}300$$
$$= 143{,}100{,}$$
$$d = 378 \text{ miles.}$$

11. (a) 7.54
     (c) 545.7
     (e) 12,300
12. The pilot will save 8.4 minutes.

Section 5.3

QUIZ

1. true	2. false	3. true	4. false	5. true
6. true	7. true	8. false	9. true	10. true

EXERCISES

3. (c)

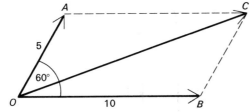

The two forces are represented by vectors $\overrightarrow{OA}$ and $\overrightarrow{OB}$. The resultant is the vector $\overrightarrow{OC}$. Note that $\angle OAC = 120°$. Therefore using the law of cosines in triangle OAC:

$$OC^2 = OA^2 + AC^2 - 2(OA)(AC) \cos 120°$$
$$= 5^2 + 10^2 + 2(5)(10) \cos 60°$$
$$= 25 + 100 + 100 \cos 60°$$
$$= 25 + 100 + 50$$
$$= 175.$$

Hence $OC = \sqrt{175} = 13.2$, approximately. In order to find the direction of $\overrightarrow{OC}$, we use the law of sines to determine $\angle AOC$:

$$\frac{\sin 120°}{13.2} = \frac{\sin (\angle AOC)}{10},$$

$$\sin(\angle AOC) = \frac{10 \sin 120°}{13.2}$$

$$= \frac{10 \sin 60°}{13.2},$$

$$\angle AOC = 41°.$$

6.

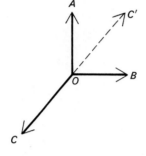

The three forces are in equilibrium, and therefore

$$\overrightarrow{OA} + \overrightarrow{OB} + \overrightarrow{OC} = \vec{0},$$

or

$$\overrightarrow{OA} + \overrightarrow{OB} = -\overrightarrow{OC} = \overrightarrow{OC'}.$$

Since $AOBC'$ is a parallelogram, $AC' = OB$ and $\angle AC'O = \angle C'OB$. In triangle AOC', by the law of sines,

(1) $$\frac{OA}{\sin \angle AC'O} = \frac{AC'}{\sin \angle AOC'}.$$

Now,

$$\frac{OA}{\sin \angle AC'O} = \frac{OA}{\sin \angle C'OB}$$

$$= \frac{OA}{\sin(180° - \angle C'OB)}$$

$$= \frac{OA}{\sin \angle BOC},$$

and

$$\frac{AC'}{\sin \angle AOC'} = \frac{OB}{\sin(180° - \angle AOC')}$$

$$= \frac{OB}{\sin \angle COA}.$$

Hence (1) becomes

$$\frac{OA}{\sin \angle BOC} = \frac{OB}{\angle COA}.$$

Similarly we can prove that

$$\frac{OA}{\sin \angle BOC} = \frac{OC}{\sin \angle AOB},$$

and the result follows.

9.

It is assumed that the effective component of the force of the wind is perpendicular to the face of the kite, and it counteracts the tension in the cord and the weight of the kite. The three forces are represented by vectors $\overrightarrow{OW}$, $\overrightarrow{OC}$, and $\overrightarrow{OA}$, respectively. Thus $\overrightarrow{OW} + \overrightarrow{OC} + \overrightarrow{OA} = \vec{0}$. We want to find the magnitude of $\overrightarrow{OA}$. Using the equation of Exercise 6,

$$\frac{OA}{\sin \angle COW} = \frac{OC}{\sin \angle AOW},$$

i.e.,

$$OA = \frac{OC \sin \angle COW}{\sin \angle AOW}$$

$$= \frac{4.5 \sin 170°}{\sin 150°}$$

$$= \frac{4.5 \sin 10°}{\sin 30°}$$

$$= 9 \sin 10°$$

$$= 1.56.$$

Hence the weight of the kite is 1.56 lbs.

CHAPTER 6

Section 6.1

QUIZ

1. true
3. true
5. true
7. true
9. true

2. true
4. true
6. true
8. false
10. false $\left(\sin (\pi/2) = 1, \text{ but } \cos 1 \neq 0.\right)$

EXERCISES

3. The cotangent function is decreasing over the intervals $(0, \pi)$ and $(\pi, 2\pi)$. It does not increase over any interval.
4. (b)

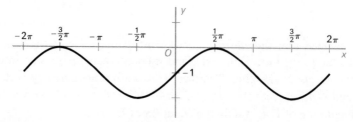

Section 6.2

QUIZ

1. true   2. false   3. false   4. false   5. true
6. true   7. false   8. true   9. false   10. false

EXERCISES

1. $y = \operatorname{cosec}^{-1} x$

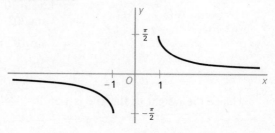

4. (b) Set $y = \cot^{-1} x$. Then

$$x = \cot y$$
$$= \frac{1}{\tan y}.$$

Hence

$$\tan y = \frac{1}{x}.$$

Therefore

$$y = \tan^{-1}\frac{1}{x},$$

since if $x > 0$ then $y = \cot^{-1} x$ is in $\left(0, \dfrac{\pi}{2}\right)$. Thus

$$\cot^{-1} x = \tan^{-1}\frac{1}{x}.$$

7. (d) Letting

$$t = 2 \sin^{-1} \tfrac{2}{3}$$

or

$$\frac{t}{2} = \sin^{-1} \tfrac{2}{3},$$

we again construct a right triangle such that $\sin \dfrac{t}{2} = \tfrac{2}{3}$.

Hence $\tan \dfrac{t}{2} = \dfrac{2}{\sqrt{5}}$, and

$$\tan (2 \sin^{-1} \tfrac{2}{3}) = \tan t$$
$$= \tan 2 \left(\frac{t}{2}\right)$$
$$= \frac{2 \tan\left(\dfrac{t}{2}\right)}{1 - \tan^2\left(\dfrac{t}{2}\right)}$$
$$= \frac{\dfrac{4}{\sqrt{5}}}{1 - \dfrac{4}{5}}$$
$$= \frac{20}{\sqrt{5}}.$$

(g) Set $\alpha = \cos^{-1}\frac{13}{14}$ and $\beta = \cos^{-1}\frac{11}{14}$. Then

$$\cos \alpha = \tfrac{13}{14}$$

and

$$\cos \beta = \tfrac{11}{14}.$$

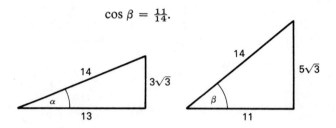

Therefore $\sin \alpha = \dfrac{3\sqrt{3}}{14}$ and $\sin \beta = \dfrac{5\sqrt{3}}{14}$. Thus

$$\sin\left(\cos^{-1}\tfrac{13}{14} + \cos^{-1}\tfrac{11}{14}\right) = \sin(\alpha + \beta)$$

$$= \sin \alpha \cos \beta + \cos \alpha \sin \beta$$

$$= \left(\frac{3\sqrt{3}}{14}\right)\left(\frac{11}{14}\right) + \left(\frac{13}{14}\right)\left(\frac{5\sqrt{3}}{14}\right)$$

$$= \frac{\sqrt{3}}{2}.$$

(j) We know that there are angles t and s such that $\tan t = 2$ and $\tan s = 3$. Hence

$$\cos(\tan^{-1} 2 + \tan^{-1} 3) = \cos(t + s)$$

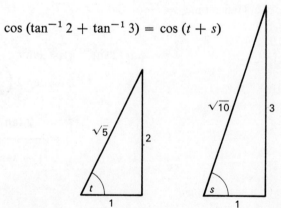

and by Theorem 1.2, Section 4.1,

$$\cos(t + s) = \cos t \cos s - \sin t \sin s$$

$$= \frac{1}{\sqrt{5}} \frac{1}{\sqrt{10}} - \frac{2}{\sqrt{5}} \frac{3}{\sqrt{10}}$$

$$= \frac{1 - 6}{5\sqrt{2}}$$

$$= -\frac{1}{\sqrt{2}}.$$

8. (a) By Theorem 2.2(f)

$$\sin^{-1} \frac{x}{\sqrt{1 + x^2}} = \tan^{-1} x.$$

By Theorems 2.1(a) and 2.2(g)

$$\sin^{-1} \frac{1}{\sqrt{1 + x^2}} = \frac{\pi}{2} - \cos^{-1} \frac{1}{\sqrt{1 + x^2}} = \frac{\pi}{2} - \tan^{-1} x.$$

Thus

$$\cos \left(\sin^{-1} \frac{x}{\sqrt{1 + x^2}} + \sin^{-1} \frac{1}{\sqrt{1 + x^2}} \right) = \cos \frac{\pi}{2} = 0.$$

Section 6.3

QUIZ

1. false	2. true	3. true	4. false	5. false
6. true	7. false	8. true	9. true	10. false

EXERCISES

3. (d) The amplitude is 1, the period is 2, the frequency is $\frac{1}{2}$, and the phase shift is 0.

   (h) The amplitude is 2, the period is 2, the frequency is $\frac{1}{2}$, and the phase shift is -1.

4. (d)

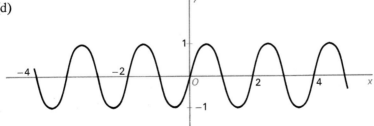

(h)

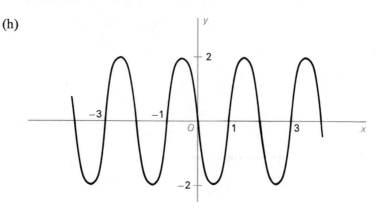

5. (d) $f(x) = \sin\left(\dfrac{\pi}{2} - x\right)$

$= \sin\left(\pi - \left(\dfrac{\pi}{2} - x\right)\right)$

$= \sin\left(x + \dfrac{\pi}{2}\right),$

where $A = 1$, $\omega = 1$, $\epsilon = \pi/2$, and the phase shift is $-\pi/2$.

(f) $f(x) = -\frac{1}{2}\sin\left(-\frac{1}{2}x - \frac{1}{2}\pi\right)$
$= \frac{1}{2}\sin\left(-\left(-\frac{1}{2}x - \frac{1}{2}\pi\right)\right)$
$= \frac{1}{2}\sin\left(\frac{1}{2}x + \frac{1}{2}\pi\right).$

Here $A = 1/2$, $\omega = 1/2$, $\epsilon = \pi/2$, and the phase shift is $-\pi$.

6. We know that

$$B\cos(\omega x + \epsilon) = B\sin\left(\dfrac{\pi}{2} - (\omega x + \epsilon)\right)$$

$$= B\sin\left(\dfrac{\pi}{2} - \omega x - \epsilon\right)$$

$$= B\sin\left(\pi - \left(\dfrac{\pi}{2} - \omega x - \epsilon\right)\right)$$

$$= B\sin\left(\omega x + \left(\epsilon + \dfrac{\pi}{2}\right)\right),$$

which is of the form (11) since $B > 0$ and $\omega > 0$. Hence the amplitude is B, the period is $(2\pi)/\omega$, and the phase shift is $-(\epsilon + \pi/2)/\omega$.

9. (b) Using the trigonometric formulas we obtain

$$-\cos x = \cos(\pi - x)$$
$$= \cos(x - \pi),$$

$$= \sin\left(\frac{\pi}{2} + x - \pi\right)$$

$$= \sin\left(x - \frac{\pi}{2}\right).$$

(d) We can write the equation as

$$f(x) = a \cos x + b \sin x,$$

where $a = -1$, $b = 1$. From Exercise 7 we can find β such that

$$\sin \beta = \frac{a}{\sqrt{a^2 + b^2}} = \frac{-1}{\sqrt{2}}$$

and

$$\cos \beta = \frac{b}{\sqrt{a^2 + b^2}} = \frac{1}{\sqrt{2}}.$$

It is obvious that $\beta = (7/4)\pi$ and hence

$$f(x) = \sqrt{2}\left(\sin\frac{7}{4}\pi \cos x + \cos\frac{7}{4}\pi \sin x\right)$$

$$= \sqrt{2} \sin\left(x + \frac{7}{4}\pi\right).$$

(g) As in part (d) we find β such that $\sin \beta = -3/5$ and $\cos \beta = 4/5$. From Table 2 we see that β is approximately $-37°$. Hence,

$$f(x) = 5 (\cos \beta \sin 2\pi x + \sin \beta \cos 2\pi x)$$
$$= 5 \sin (\beta + 2\pi x)$$
$$= 5 \sin (2\pi x - 37°).$$

CHAPTER 7

Section 7.1

QUIZ

1. true	2. true	3. false	4. false	5. true
6. false	7. true	8. true	9. false	10. false

EXERCISES

1. (e) If $z = 1 + i$, then $\bar{z} = 1 - i$, $|z| = \sqrt{2}$.
   (o) Here

$$(\sqrt{3} + i) + (2\sqrt{3} - 3i) = 3\sqrt{3} - 2i;$$

the complex conjugate is $3\sqrt{3} + 2i$, and the modulus is $\sqrt{31}$.

2. (i) $\dfrac{1 + 2i}{2 - i} + \dfrac{2i}{-3 + i} = \dfrac{(1 + 2i)(2 + i)}{(2 - i)(2 + i)} + \dfrac{2i(-3 - i)}{(-3 + i)(-3 - i)}$

$$= \frac{5i}{5} + \frac{2 - 6i}{10}$$

$$= \frac{10i + 2 - 6i}{10}$$

$$= \frac{2 + 4i}{10}$$

$$= \frac{1}{5} + \frac{2}{5} i.$$

(p) $\dfrac{(2 - i)(1 + 3i)}{(1 + i)(3 + 2i)} = \dfrac{5 + 5i}{1 + 5i}$

$$= \frac{(5 + 5i)(1 - 5i)}{(1 + 5i)(1 - 5i)}$$

$$= \frac{30 - 20i}{26}$$

$$= \frac{15}{13} - \frac{10}{13} i.$$

3. (a) Re(z) = 0 if and only if $a = 0$, where $z = a + ib$. But $a = (z + \bar{z})/2$, and hence $a = 0$ if and only if $\bar{z} = -z$.
   (b) Im(z) = 0 if and only if $b = 0$, where $z = a + ib$. But $b = (z - \bar{z})/2$, and hence $b = 0$ if and only if $z = \bar{z}$.
   (c) We know that $|z| = 0$ if and only if $z = 0$. Hence $|z_1 z_2| = 0$ if and only if $z_1 z_2 = 0$, i.e., if and only if $z_1 = 0$ or $z_2 = 0$.

7. (h) First we must simplify this expression to the form $x + yi$:

$$\left(-\frac{1}{2} + \frac{\sqrt{3}}{2} i\right)^2 = -\frac{1}{2} - \frac{\sqrt{3}}{2} i.$$

Using formula (9) we obtain

$$z^{-1} = \frac{-\frac{1}{2}}{\frac{1}{4} + \frac{3}{4}} + \frac{-\left(-\frac{\sqrt{3}}{2}\right)}{\frac{1}{4} + \frac{3}{4}} i = -\frac{1}{2} + \frac{\sqrt{3}}{2} i.$$

Alternatively,

$$z^{-1} = \frac{1}{-\frac{1}{2} - \frac{\sqrt{3}}{2} i} = \frac{-\frac{1}{2} + \frac{\sqrt{3}}{2} i}{\left(-\frac{1}{2} - \frac{\sqrt{3}}{2} i\right)\left(-\frac{1}{2} + \frac{\sqrt{3}}{2} i\right)} = -\frac{1}{2} + \frac{\sqrt{3}}{2} i.$$

Section 7.2

QUIZ

1. false	2. true	3. false	4. true	5. false
6. false	7. true	8. false	9. false	10. true

EXERCISES

2.

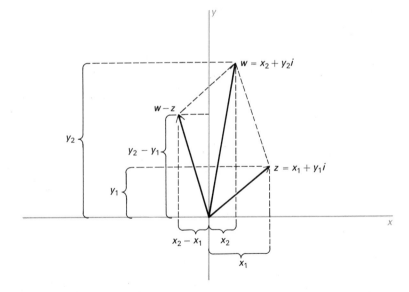

3. (b) $|z| = |w|$ means that the complex numbers z and w are the same distance from the origin in the Argand diagram.

(c) $\text{Re}(z) = \text{Im}(z)$ means that $a = b$, where $z = a + ib$. Hence the complex number z must lie on the line through the origin inclined at 45° to the positive horizontal axis.

(h) $\text{Re}(z) < 0$, $\text{Im}(z) < 0$ means that z lies in the third quadrant.

(i) $\text{Re}(z) < 0$, $\text{Im}(z) > 0$ means that z lies in the second quadrant.

4. (d) $(r, \theta) = (\pi, \pi)$,

$$x = r \cos \theta$$
$$= \pi \cos \pi$$
$$= -\pi,$$
$$y = r \sin \theta$$
$$= \pi \sin \pi$$
$$= 0.$$

Thus $(x, y) = (-\pi, 0)$.

(f) $(r, \theta) = (2, 135°)$

$$= \left(2, \frac{3\pi}{4}\right),$$

$$x = r \cos \theta$$
$$= 2 \cos \frac{3\pi}{4}$$
$$= \frac{-2}{\sqrt{2}}$$
$$= -\sqrt{2},$$
$$y = r \sin \theta$$
$$= 2 \sin \frac{3\pi}{4}$$
$$= \frac{2}{\sqrt{2}}$$
$$= \sqrt{2}.$$

Thus $(x, y) = (-\sqrt{2}, \sqrt{2})$.

5. (g)   We have

$$\text{amp}(1 - i\sqrt{3}) = -60°,$$
$$\text{amp}(1 - i) = -45°,$$
$$\text{amp}(2i(\sqrt{3} - 1)) = 90°.$$

Thus, by Theorem 2.4

$$\text{amp} \frac{(1 - i\sqrt{3})(1 - i)}{2i(\sqrt{3} - 1)} = -60° - 45° - 90°$$

$$= -195°$$

and

$$\text{amp} \left(\frac{(1 - i\sqrt{3})(1 - i)}{2i(\sqrt{3} - 1)}\right)^4 = -780° \quad \text{or} \quad -60°.$$

Also,

$$|1 - i\sqrt{3}| = 2,$$
$$|1 - i| = \sqrt{2},$$
$$|2i(\sqrt{3} - 1)| = 2(\sqrt{3} - 1).$$

Hence

$$\left|\left(\frac{(1 - i\sqrt{3})(1 - i)}{2i(\sqrt{3} - 1)}\right)^4\right| = \left(\frac{2\sqrt{2}}{2(\sqrt{3} - 1)}\right)^4$$

$$= \left(\frac{\sqrt{2}(\sqrt{3} + 1)}{3 - 1}\right)^4$$

$$= \frac{(\sqrt{3} + 1)^4}{4}$$

$$= 7 + 4\sqrt{3}.$$

Thus the required polar form is

$$(7 + 4\sqrt{3})(\cos(-60°) + i\sin(-60°)).$$

6. (a) $w_k = \left(\cos\dfrac{0° + k \cdot 360°}{n} + i\sin\dfrac{0° + k \cdot 360°}{n}\right), k = 0, 1, \ldots, n - 1.$

10. (b) In this case, $|z| = \sqrt{0 + 1} = 1$, and $\text{amp}(z) = 270°$. The polar form of z is therefore

$$(\cos 270° + i\sin 270°),$$

and hence, using Theorem 2.4, the 6th roots of z are

$$w_k = \left(\cos\frac{270° + k \cdot 360°}{6} + i\sin\frac{270° + k \cdot 360°}{6}\right),$$

$k = 0, 1, 2, 3, 4, 5.$

(d) Here

$$|z| = \sqrt{4 + 1}$$
$$= \sqrt{5},$$

and

$$\text{amp}(z) = \tan^{-1}\tfrac{1}{2}$$
$$= 26°30'.$$

The polar form of z is

$$\sqrt{5}(\cos 26°30' + i\sin 26°30').$$

We know from Theorem 2.4, the 5th roots of z are

$$w_k = (\sqrt{5})^{1/5} \left(\cos \frac{26°30' + k \cdot 360°}{5} + i \sin \frac{26°30' + k \cdot 360°}{5} \right),$$

$k = 0, 1, 2, 3, 4.$

Tables

Table 1 Common Logarithms

x	0	1	2	3	4	5	6	7	8	9
1.0	.0000	.0043	.0086	.0128	.0170	.0212	.0253	.0294	.0334	.0374
1.1	.0414	.0453	.0492	.0531	.0569	.0607	.0645	.0682	.0719	.0755
1.2	.0792	.0828	.0864	.0899	.0934	.0969	.1004	.1038	.1072	.1106
1.3	.1139	.1173	.1206	.1239	.1271	.1303	.1335	.1367	.1399	.1430
1.4	.1461	.1492	.1523	.1553	.1584	.1614	.1644	.1673	.1703	.1732
1.5	.1761	.1790	.1818	.1847	.1875	.1903	.1931	.1959	.1987	.2014
1.6	.2041	.2068	.2095	.2122	.2148	.2175	.2201	.2227	.2253	.2279
1.7	.2304	.2330	.2355	.2380	.2405	.2430	.2455	.2480	.2504	.2529
1.8	.2553	.2577	.2601	.2625	.2648	.2672	.2695	.2718	.2742	.2765
1.9	.2788	.2810	.2833	.2856	.2878	.2900	.2923	.2945	.2967	.2989
2.0	.3010	.3032	.3054	.3075	.3096	.3118	.3139	.3160	.3181	.3201
2.1	.3222	.3243	.3263	.3284	.3304	.3324	.3345	.3365	.3385	.3404
2.2	.3424	.3444	.3464	.3483	.3502	.3522	.3541	.3560	.3579	.3598
2.3	.3617	.3636	.3655	.3674	.3692	.3711	.3729	.3747	.3766	.3784
2.4	.3802	.3820	.3838	.3856	.3874	.3892	.3909	.3927	.3945	.3962
2.5	.3979	.3997	.4014	.4031	.4048	.4065	.4082	.4099	.4116	.4133
2.6	.4150	.4166	.4183	.4200	.4216	.4232	.4249	.4265	.4281	.4298
2.7	.4314	.4330	.4346	.4362	.4378	.4393	.4409	.4425	.4440	.4456
2.8	.4472	.4487	.4502	.4518	.4533	.4548	.4564	.4579	.4594	.4609
2.9	.4624	.4639	.4654	.4669	.4683	.4698	.4713	.4728	.4742	.4757
3.0	.4771	.4786	.4800	.4814	.4829	.4843	.4857	.4871	.4886	.4900
3.1	.4914	.4928	.4942	.4955	.4969	.4983	.4997	.5011	.5024	.5038
3.2	.5051	.5065	.5079	.5092	.5105	.5119	.5132	.5145	.5159	.5172
3.3	.5185	.5198	.5211	.5224	.5237	.5250	.5263	.5276	.5289	.5302
3.4	.5315	.5328	.5340	.5353	.5366	.5378	.5391	.5403	.5416	.5428
3.5	.5441	.5453	.5465	.5478	.5490	.5502	.5514	.5527	.5539	.5551
3.6	.5563	.5575	.5587	.5599	.5611	.5623	.5635	.5647	.5658	.5670
3.7	.5682	.5694	.5705	.5717	.5729	.5740	.5752	.5763	.5775	.5786
3.8	.5798	.5809	.5821	.5832	.5843	.5855	.5866	.5877	.5888	.5899
3.9	.5911	.5922	.5933	.5944	.5955	.5966	.5977	.5988	.5999	.6010
4.0	.6021	.6031	.6042	.6053	.6064	.6075	.6085	.6096	.6107	.6117
4.1	.6128	.6138	.6149	.6160	.6170	.6180	.6191	.6201	.6212	.6222
4.2	.6232	.6243	.6253	.6263	.6274	.6284	.6294	.6304	.6314	.6325
4.3	.6335	.6345	.6355	.6365	.6375	.6385	.6395	.6405	.6415	.6425
4.4	.6435	.6444	.6454	.6464	.6474	.6484	.6493	.6503	.6513	.6522
4.5	.6532	.6542	.6551	.6561	.6571	.6580	.6590	.6599	.6609	.6618
4.6	.6628	.6637	.6646	.6656	.6665	.6675	.6684	.6693	.6702	.6712
4.7	.6721	.6730	.6739	.6749	.6758	.6767	.6776	.6785	.6794	.6803
4.8	.6812	.6821	.6830	.6839	.6848	.6857	.6866	.6875	.6884	.6893
4.9	.6902	.6911	.6920	.6928	.6937	.6946	.6955	.6964	.6972	.6981
5.0	.6990	.6998	.7007	.7016	.7024	.7033	.7042	.7050	.7059	.7067
5.1	.7076	.7084	.7093	.7101	.7110	.7118	.7126	.7135	.7143	.7152
5.2	.7160	.7168	.7177	.7185	.7193	.7202	.7210	.7218	.7226	.7235
5.3	.7243	.7251	.7259	.7267	.7275	.7284	.7292	.7300	.7308	.7316
5.4	.7324	.7332	.7340	.7348	.7356	.7364	.7372	.7380	.7388	.7396
x	0	1	2	3	4	5	6	7	8	9

Table 1   Common Logarithms

x	0	1	2	3	4	5	6	7	8	9
5.5	.7404	.7412	.7419	.7427	.7435	.7443	.7451	.7459	.7466	.7474
5.6	.7482	.7490	.7497	.7505	.7513	.7520	.7528	.7536	.7543	.7551
5.7	.7559	.7566	.7574	.7582	.7589	.7597	.7604	.7612	.7619	.7627
5.8	.7634	.7642	.7649	.7657	.7664	.7672	.7679	.7686	.7694	.7701
5.9	.7709	.7716	.7723	.7731	.7738	.7745	.7752	.7760	.7767	.7774
6.0	.7782	.7789	.7796	.7803	.7810	.7818	.7825	.7832	.7839	.7846
6.1	.7853	.7860	.7868	.7875	.7882	.7889	.7896	.7903	.7910	.7917
6.2	.7924	.7931	.7938	.7945	.7952	.7959	.7966	.7973	.7980	.7987
6.3	.7993	.8000	.8007	.8014	.8021	.8028	.8035	.8041	.8048	.8055
6.4	.8062	.8069	.8075	.8082	.8089	.8096	.8102	.8109	.8116	.8122
6.5	.8129	.8136	.8142	.8149	.8156	.8162	.8169	.8176	.8182	.8189
6.6	.8195	.8202	.8209	.8215	.8222	.8228	.8235	.8241	.8248	.8254
6.7	.8261	.8267	.8274	.8280	.8287	.8293	.8299	.8306	.8312	.8319
6.8	.8325	.8331	.8338	.8344	.8351	.8357	.8363	.8370	.8376	.8382
6.9	.8388	.8395	.8401	.8407	.8414	.8420	.8426	.8432	.8439	.8445
7.0	.8451	.8457	.8463	.8470	.8476	.8482	.8488	.8494	.8500	.8506
7.1	.8513	.8519	.8525	.8531	.8537	.8543	.8549	.8555	.8561	.8567
7.2	.8573	.8579	.8585	.8591	.8597	.8603	.8609	.8615	.8621	.8627
7.3	.8633	.8639	.8645	.8651	.8657	.8663	.8669	.8675	.8681	.8686
7.4	.8692	.8698	.8704	.8710	.8716	.8722	.8727	.8733	.8739	.8745
7.5	.8751	.8756	.8762	.8768	.8774	.8779	.8785	.8791	.8797	.8802
7.6	.8808	.8814	.8820	.8825	.8831	.8837	.8842	.8848	.8854	.8859
7.7	.8865	.8871	.8876	.8882	.8887	.8893	.8899	.8904	.8910	.8915
7.8	.8921	.8927	.8932	.8938	.8943	.8949	.8954	.8960	.8965	.8971
7.9	.8976	.8982	.8987	.8993	.8998	.9004	.9009	.9015	.9020	.9025
8.0	.9031	.9036	.9042	.9047	.9053	.9058	.9063	.9069	.9074	.9079
8.1	.9085	.9090	.9096	.9101	.9106	.9112	.9117	.9122	.9128	.9133
8.2	.9138	.9143	.9149	.9154	.9159	.9165	.9170	.9175	.9180	.9186
8.3	.9191	.9196	.9201	.9206	.9212	.9217	.9222	.9227	.9232	.9238
8.4	.9243	.9248	.9253	.9258	.9263	.9269	.9274	.9279	.9284	.9289
8.5	.9294	.9299	.9304	.9309	.9315	.9320	.9325	.9330	.9335	.9340
8.6	.9345	.9350	.9355	.9360	.9365	.9370	.9375	.9380	.9385	.9390
8.7	.9395	.9400	.9405	.9410	.9415	.9420	.9425	.9430	.9435	.9440
8.8	.9445	.9450	.9455	.9460	.9465	.9469	.9474	.9479	.9484	.9489
8.9	.9494	.9499	.9504	.9509	.9513	.9518	.9523	.9528	.9533	.9538
9.0	.9542	.9547	.9552	.9557	.9562	.9566	.9571	.9576	.9581	.9586
9.1	.9590	.9595	.9600	.9605	.9609	.9614	.9619	.9624	.9628	.9633
9.2	.9638	.9643	.9647	.9652	.9657	.9661	.9666	.9671	.9675	.9680
9.3	.9685	.9689	.9694	.9699	.9703	.9708	.9713	.9717	.9722	.9727
9.4	.9731	.9736	.9741	.9745	.9750	.9754	.9759	.9763	.9768	.9773
9.5	.9777	.9782	.9786	.9791	.9795	.9800	.9805	.9809	.9814	.9818
9.6	.9823	.9827	.9832	.9836	.9841	.9845	.9850	.9854	.9859	.9863
9.7	.9868	.9872	.9877	.9881	.9886	.9890	.9894	.9899	.9903	.9908
9.8	.9912	.9917	.9921	.9926	.9930	.9934	.9939	.9943	.9948	.9952
9.9	.9956	.9961	.9965	.9969	.9974	.9978	.9983	.9987	.9991	.9996
x	0	1	2	3	4	5	6	7	8	9

Table 2   Trigonometric Functions

Degrees	Radians	Sine	Cosine	Tangent
0°0′	.0000	.0000	1.000	.0000
0°30′	.0087	.0087	1.000	.0087
1°0′	.0175	.0174	.9999	.0174
1°30′	.0262	.0261	.9997	.0262
2°0′	.0349	.0349	.9994	.0349
2°30′	.0436	.0436	.9991	.0437
3°0′	.0524	.0523	.9986	.0524
3°30′	.0611	.0611	.9981	.0612
4°0′	.0698	.0698	.9976	.0699
4°30′	.0785	.0785	.9969	.0787
5°0′	.0873	.0872	.9962	.0875
5°30′	.0960	.0959	.9954	.0963
6°0′	.1047	.1045	.9945	.1051
6°30′	.1135	.1132	.9936	.1139
7°0′	.1222	.1219	.9926	.1228
7°30′	.1309	.1305	.9914	.1317
8°0′	.1396	.1392	.9903	.1405
8°30′	.1484	.1478	.9890	.1495
9°0′	.1571	.1564	.9877	.1584
9°30′	.1658	.1651	.9863	.1673
10°0′	.1745	.1737	.9848	.1763
10°30′	.1833	.1822	.9833	.1853
11°0′	.1920	.1908	.9816	.1944
11°30′	.2007	.1994	.9799	.2035
12°0′	.2094	.2079	.9782	.2126
12°30′	.2182	.2164	.9763	.2217
13°0′	.2269	.2250	.9744	.2309
13°30′	.2356	.2335	.9724	.2401
14°0′	.2444	.2419	.9703	.2493
14°30′	.2531	.2504	.9682	.2586
15°0′	.2618	.2588	.9659	.2679
15°30′	.2705	.2672	.9636	.2773
16°0′	.2793	.2756	.9613	.2868
16°30′	.2880	.2840	.9588	.2962
17°0′	.2967	.2924	.9563	.3057
17°30′	.3054	.3007	.9537	.3153
18°0′	.3142	.3090	.9511	.3249
18°30′	.3229	.3173	.9483	.3346
19°0′	.3316	.3256	.9455	.3443
19°30′	.3403	.3338	.9426	.3541

Table 2   Trigonometric Functions

Degrees	Radians	Sine	Cosine	Tangent
20°0′	.3491	.3420	.9397	.3640
20°30′	.3578	.3502	.9367	.3739
21°0′	.3665	.3584	.9336	.3839
21°30′	.3753	.3665	.9304	.3939
22°0′	.3840	.3746	.9272	.4040
22°30′	.3927	.3827	.9239	.4142
23°0′	.4014	.3907	.9205	.4245
23°30′	.4102	.3988	.9171	.4348
24°0′	.4189	.4067	.9136	.4452
24°30′	.4276	.4147	.9100	.4557
25°0′	.4363	.4226	.9063	.4663
25°30′	.4451	.4305	.9026	.4770
26°0′	.4538	.4384	.8988	.4877
26°30′	.4625	.4462	.8949	.4986
27°0′	.4712	.4540	.8910	.5095
27°30′	.4800	.4618	.8870	.5206
28°0′	.4887	.4695	.8830	.5317
28°30′	.4974	.4772	.8788	.5430
29°0′	.5061	.4848	.8746	.5543
29°30′	.5149	.4924	.8704	.5658
30°0′	.5236	.5000	.8660	.5774
30°30′	.5323	.5075	.8616	.5891
31°0′	.5411	.5150	.8572	.6009
31°30′	.5498	.5225	.8526	.6128
32°0′	.5585	.5299	.8481	.6249
32°30′	.5672	.5373	.8434	.6371
33°0′	.5760	.5446	.8387	.6494
33°30′	.5847	.5519	.8339	.6619
34°0′	.5934	.5592	.8290	.6745
34°30′	.6021	.5664	.8241	.6873
35°0′	.6109	.5736	.8192	.7002
35°30′	.6196	.5807	.8141	.7133
36°0′	.6283	.5878	.8090	.7265
36°30′	.6371	.5948	.8039	.7400
37°0′	.6458	.6018	.7986	.7536
37°30′	.6545	.6088	.7934	.7673
38°0′	.6632	.6157	.7880	.7813
38°30′	.6720	.6225	.7826	.7954
39°0′	.6807	.6293	.7772	.8098
39°30′	.6894	.6361	.7716	.8243

Table 2 Trigonometric Functions

Degrees	Radians	Sine	Cosine	Tangent
40°0′	.6981	.6428	.7660	.8391
40°30′	.7069	.6495	.7604	.8541
41°0′	.7156	.6561	.7547	.8693
41°30′	.7243	.6626	.7490	.8847
42°0′	.7330	.6691	.7431	.9004
42°30′	.7418	.6756	.7373	.9163
43°0′	.7505	.6820	.7314	.9325
43°30′	.7592	.6884	.7254	.9490
44°0′	.7679	.6947	.7193	.9657
44°30′	.7767	.7009	.7133	.9827
45°0′	.7854	.7071	.7071	1.000
45°30′	.7941	.7133	.7009	1.018
46°0′	.8029	.7193	.6947	1.036
46°30′	.8116	.7254	.6884	1.054
47°0′	.8203	.7314	.6820	1.072
47°30′	.8290	.7373	.6756	1.091
48°0′	.8378	.7431	.6691	1.111
48°30′	.8465	.7490	.6626	1.130
49°0′	.8552	.7547	.6561	1.150
49°30′	.8639	.7604	.6495	1.171
50°0′	.8727	.7660	.6428	1.192
50°30′	.8814	.7716	.6361	1.213
51°0′	.8901	.7771	.6293	1.235
51°30′	.8988	.7826	.6225	1.257
52°0′	.9076	.7880	.6157	1.280
52°30′	.9163	.7934	.6088	1.303
53°0′	.9250	.7986	.6018	1.327
53°30′	.9338	.8039	.5948	1.351
54°0′	.9425	.8090	.5878	1.376
54°30′	.9512	.8141	.5807	1.401
55°0′	.9599	.8192	.5736	1.428
55°30′	.9687	.8241	.5664	1.455
56°0′	.9774	.8290	.5592	1.483
56°30′	.9861	.8339	.5519	1.511
57°0′	.9948	.8387	.5446	1.540
57°30′	1.004	.8434	.5373	1.570
58°0′	1.012	.8481	.5299	1.600
58°30′	1.021	.8526	.5225	1.632
59°0′	1.030	.8572	.5150	1.664
59°30′	1.038	.8616	.5075	1.698

Table 2   Trigonometric Functions

Degrees	Radians	Sine	Cosine	Tangent
60°0′	1.047	.8660	.5000	1.732
60°30′	1.056	.8704	.4924	1.768
61°0′	1.065	.8746	.4848	1.804
61°30′	1.073	.8788	.4772	1.842
62°0′	1.082	.8830	.4695	1.881
62°30′	1.091	.8870	.4618	1.921
63°0′	1.100	.8910	.4540	1.963
63°30′	1.108	.8949	.4462	2.006
64°0′	1.117	.8988	.4384	2.050
64°30′	1.126	.9026	.4305	2.097
65°0′	1.134	.9063	.4226	2.145
65°30′	1.143	.9100	.4147	2.194
66°0′	1.152	.9136	.4067	2.246
66°30′	1.161	.9171	.3988	2.300
67°0′	1.169	.9205	.3907	2.356
67°30′	1.178	.9239	.3827	2.414
68°0′	1.187	.9272	.3746	2.475
68°30′	1.196	.9304	.3665	2.539
69°0′	1.204	.9336	.3584	2.605
69°30′	1.213	.9367	.3502	2.675
70°0′	1.222	.9397	.3420	2.748
70°30′	1.230	.9426	.3338	2.824
71°0′	1.239	.9455	.3256	2.904
71°30′	1.248	.9483	.3173	2.989
72°0′	1.257	.9511	.3090	3.078
72°30′	1.265	.9537	.3007	3.172
73°0′	1.274	.9563	.2924	3.271
73°30′	1.283	.9588	.2840	3.376
74°0′	1.292	.9613	.2756	3.487
74°30′	1.300	.9636	.2672	3.606
75°0′	1.309	.9659	.2588	3.732
75°30′	1.318	.9682	.2504	3.867
76°0′	1.326	.9703	.2419	4.011
76°30′	1.335	.9724	.2335	4.165
77°0′	1.344	.9744	.2250	4.332
77°30′	1.353	.9663	.2164	4.511
78°0′	1.361	.9782	.2079	4.705
78°30′	1.370	.9799	.1994	4.915
79°0′	1.379	.9816	.1908	5.145
79°30′	1.388	.9833	.1822	5.396

Table 2   Trigonometric Functions

Degrees	Radians	Sine	Cosine	Tangent
80°0′	1.396	.9848	.1737	5.671
80°30′	1.405	.9863	.1651	5.976
81°0′	1.414	.9877	.1564	6.314
81°30′	1.422	.9890	.1478	6.691
82°0′	1.431	.9903	.1392	7.115
82°30′	1.440	.9914	.1305	7.596
83°0′	1.449	.9926	.1219	8.144
83°30′	1.457	.9936	.1132	8.777
84°0′	1.466	.9945	.1045	9.514
84°30′	1.475	.9954	.0959	10.39
85°0′	1.484	.9962	.0872	11.43
85°30′	1.492	.9969	.0785	12.71
86°0′	1.501	.9976	.0698	14.30
86°30′	1.510	.9981	.0611	16.35
87°0′	1.518	.9986	.0523	19.08
87°30′	1.527	.9991	.0436	22.90
88°0′	1.536	.9994	.0349	28.64
88°30′	1.545	.9997	.0262	38.19
89°0′	1.553	.9999	.0175	57.29
89°30′	1.562	1.000	.0087	114.6
90°0′	1.571	1.000	.0000	undefined

Table 3   Logs of the Trigonometric Functions

Degrees	Radians	Log of Sin	Log of Cos	Log of Tan
0°0′	.0000	undefined	.0000	undefined
0°30′	.0087	.9408 − 3	.0000	.9408 − 3
1°0′	.0175	.2419 − 2	.9999 − 1	.2419 − 2
1°30′	.0262	.4179 − 2	.9999 − 1	.4180 − 2
2°0′	.0349	.5428 − 2	.9997 − 1	.5430 − 2
2°30′	.0436	.6397 − 2	.9996 − 1	.6401 − 2
3°0′	.0524	.7188 − 2	.9994 − 1	.7194 − 2
3°30′	.0611	.7857 − 2	.9992 − 1	.7865 − 2
4°0′	.0698	.8436 − 2	.9989 − 1	.8446 − 2
4°30′	.0785	.8946 − 2	.9987 − 1	.8960 − 2
5°0′	.0873	.9403 − 2	.9983 − 1	.9420 − 2
5°30′	.0960	.9816 − 2	.9980 − 1	.9836 − 2
6°0′	.1047	.0192 − 1	.9976 − 1	.0216 − 2
6°30′	.1135	.0539 − 1	.9972 − 1	.0567 − 1
7°0′	.1222	.0859 − 1	.9968 − 1	.0891 − 1
7°30′	.1309	.1157 − 1	.9963 − 1	.1194 − 1
8°0′	.1396	.1436 − 1	.9958 − 1	.1478 − 1
8°30′	.1484	.1697 − 1	.9952 − 1	.1745 − 1
9°0′	.1571	.1943 − 1	.9946 − 1	.1997 − 1
9°30′	.1658	.2176 − 1	.9940 − 1	.2236 − 1
10°0′	.1745	.2397 − 1	.9934 − 1	.2463 − 1
10°30′	.1833	.2606 − 1	.9927 − 1	.2680 − 1
11°0′	.1920	.2806 − 1	.9920 − 1	.2887 − 1
11°30′	.2007	.2997 − 1	.9912 − 1	.3085 − 1
12°0′	.2094	.3179 − 1	.9904 − 1	.3275 − 1
12°30′	.2182	.3353 − 1	.9896 − 1	.3458 − 1
13°0′	.2269	.3521 − 1	.9887 − 1	.3634 − 1
13°30′	.2356	.3682 − 1	.9878 − 1	.3804 − 1
14°0′	.2444	.3837 − 1	.9869 − 1	.3968 − 1
14°30′	.2531	.3986 − 1	.9859 − 1	.4127 − 1
15°0′	.2618	.4130 − 1	.9849 − 1	.4281 − 1
15°30′	.2705	.4269 − 1	.9839 − 1	.4430 − 1
16°0′	.2793	.4403 − 1	.9828 − 1	.4575 − 1
16°30′	.2880	.4533 − 1	.9817 − 1	.4716 − 1
17°0′	.2967	.4659 − 1	.9806 − 1	.4853 − 1
17°30′	.3054	.4781 − 1	.9794 − 1	.4987 − 1
18°0′	.3142	.4900 − 1	.9782 − 1	.5118 − 1
18°30′	.3229	.5015 − 1	.9770 − 1	.5245 − 1
19°0′	.3316	.5126 − 1	.9757 − 1	.5370 − 1
19°30′	.3403	.5235 − 1	.9744 − 1	.5492 − 1

Table 3 Logs of the Trigonometric Functions

Degrees	Radians	Log of Sin	Log of Cos	Log of Tan
20°0′	.3491	.5341 − 1	.9730 − 1	.5611 − 1
20°30′	.3578	.5443 − 1	.9716 − 1	.5727 − 1
21°0′	.3665	.5543 − 1	.9702 − 1	.5842 − 1
21°30′	.3753	.5641 − 1	.9687 − 1	.5954 − 1
22°0′	.3840	.5736 − 1	.9672 − 1	.6064 − 1
22°30′	.3927	.5828 − 1	.9656 − 1	.6172 − 1
23°0′	.4014	.5919 − 1	.9640 − 1	.6279 − 1
23°30′	.4102	.6007 − 1	.9624 − 1	.6383 − 1
24°0′	.4189	.6093 − 1	.9607 − 1	.6486 − 1
24°30′	.4276	.6177 − 1	.9590 − 1	.6587 − 1
25°0′	.4363	.6260 − 1	.9573 − 1	.6687 − 1
25°30′	.4451	.6340 − 1	.9555 − 1	.6785 − 1
26°0′	.4538	.6418 − 1	.9537 − 1	.6882 − 1
26°30′	.4625	.6495 − 1	.9518 − 1	.6977 − 1
27°0′	.4712	.6571 − 1	.9499 − 1	.7072 − 1
27°30′	.4800	.6644 − 1	.9479 − 1	.7165 − 1
28°0′	.4887	.6716 − 1	.9459 − 1	.7257 − 1
28°30′	.4974	.6787 − 1	.9439 − 1	.7348 − 1
29°0′	.5062	.6856 − 1	.9418 − 1	.7438 − 1
29°30′	.5149	.6923 − 1	.9397 − 1	.7526 − 1
30°0′	.5236	.6990 − 1	.9375 − 1	.7614 − 1
30°30′	.5323	.7055 − 1	.9353 − 1	.7702 − 1
31°0′	.5411	.7118 − 1	.9331 − 1	.7788 − 1
31°30′	.5498	.7181 − 1	.9308 − 1	.7873 − 1
32°0′	.5585	.7242 − 1	.9284 − 1	.7958 − 1
32°30′	.5672	.7302 − 1	.9260 − 1	.8042 − 1
33°0′	.5760	.7361 − 1	.9236 − 1	.8125 − 1
33°30′	.5847	.7419 − 1	.9211 − 1	.8208 − 1
34°0′	.5934	.7476 − 1	.9186 − 1	.8290 − 1
34°30′	.6021	.7531 − 1	.9160 − 1	.8371 − 1
35°0′	.6109	.7586 − 1	.9134 − 1	.8452 − 1
35°30′	.6196	.7640 − 1	.9107 − 1	.8533 − 1
36°0′	.6283	.7692 − 1	.9080 − 1	.8613 − 1
36°30′	.6371	.7744 − 1	.9052 − 1	.8692 − 1
37°0′	.6458	.7795 − 1	.9024 − 1	.8771 − 1
37°30′	.6545	.7845 − 1	.8995 − 1	.8850 − 1
38°0′	.6632	.7893 − 1	.8965 − 1	.8928 − 1
38°30′	.6720	.7942 − 1	.8935 − 1	.9006 − 1
39°0′	.6807	.7989 − 1	.8905 − 1	.9084 − 1
39°30′	.6894	.8035 − 1	.8874 − 1	.9161 − 1

Table 3   Logs of the Trigonometric Functions

Degrees	Radians	Log of Sin	Log of Cos	Log of Tan
40°0′	.6981	.8081 − 1	.8843 − 1	.9238 − 1
40°30′	.7069	.8125 − 1	.8811 − 1	.9315 − 1
41°0′	.7156	.8169 − 1	.8778 − 1	.9392 − 1
41°30′	.7243	.8213 − 1	.8745 − 1	.9468 − 1
42°0′	.7330	.8255 − 1	.8711 − 1	.9544 − 1
42°30′	.7418	.8297 − 1	.8676 − 1	.9621 − 1
43°0′	.7505	.8338 − 1	.8641 − 1	.9697 − 1
43°30′	.7592	.8378 − 1	.8606 − 1	.9773 − 1
44°0′	.7679	.8418 − 1	.8569 − 1	.9848 − 1
44°30′	.7767	.8457 − 1	.8532 − 1	.9924 − 1
45°0′	.7854	.8495 − 1	.8495 − 1	.0000
45°30′	.7941	.8532 − 1	.8457 − 1	.0076
46°0′	.8029	.8569 − 1	.8418 − 1	.0152
46°30′	.8116	.8601 − 1	.8378 − 1	.0228
47°0′	.8203	.8641 − 1	.8338 − 1	.0303
47°30′	.8290	.8676 − 1	.8297 − 1	.0380
48°0′	.8378	.8711 − 1	.8255 − 1	.0456
48°30′	.8465	.8745 − 1	.8213 − 1	.0532
49°0′	.8552	.8778 − 1	.8169 − 1	.0608
49°30′	.8639	.8811 − 1	.8125 − 1	.0685
50°0′	.8727	.8843 − 1	.8081 − 1	.0762
50°30′	.8814	.8874 − 1	.8035 − 1	.0839
51°0′	.8901	.8905 − 1	.7989 − 1	.0916
51°30′	.8988	.8935 − 1	.7942 − 1	.0994
52°0′	.9076	.8965 − 1	.7893 − 1	.1072
52°30′	.9163	.8995 − 1	.7845 − 1	.1150
53°0′	.9250	.9024 − 1	.7795 − 1	.1229
53°30′	.9338	.9052 − 1	.7744 − 1	.1308
54°0′	.9425	.9080 − 1	.7692 − 1	.1387
54°30′	.9512	.9107 − 1	.7640 − 1	.1467
55°0′	.9599	.9134 − 1	.7586 − 1	.1548
55°30′	.9687	.9160 − 1	.7531 − 1	.1629
56°0′	.9774	.9186 − 1	.7476 − 1	.1710
56°30′	.9861	.9211 − 1	.7419 − 1	.1792
57°0′	.9948	.9236 − 1	.7361 − 1	.1875
57°30′	1.004	.9260 − 1	.7302 − 1	.1958
58°0′	1.012	.9284 − 1	.7242 − 1	.2042
58°30′	1.021	.9308 − 1	.7181 − 1	.2127
59°0′	1.030	.9331 − 1	.7118 − 1	.2212
59°30′	1.038	.9353 − 1	.7055 − 1	.2299

Table 3   Logs of the Trigonometric Functions

Degrees	Radians	Log of Sin	Log of Cos	Log of Tan
60°0′	1.047	.9375 − 1	.6990 − 1	.2386
60°30′	1.056	.9397 − 1	.6923 − 1	.2474
61°0′	1.065	.9418 − 1	.6856 − 1	.2563
61°30′	1.073	.9439 − 1	.6787 − 1	.2652
62°0′	1.082	.9459 − 1	.6716 − 1	.2743
62°30′	1.091	.9479 − 1	.6644 − 1	.2835
63°0′	1.100	.9499 − 1	.6571 − 1	.2928
63°30′	1.108	.9518 − 1	.6495 − 1	.3023
64°0′	1.117	.9537 − 1	.6418 − 1	.3118
64°30′	1.126	.9555 − 1	.6340 − 1	.3215
65°0′	1.134	.9573 − 1	.6260 − 1	.3313
65°30′	1.143	.9590 − 1	.6177 − 1	.3413
66°0′	1.152	.9607 − 1	.6093 − 1	.3514
66°30′	1.161	.9624 − 1	.6007 − 1	.3617
67°0′	1.169	.9640 − 1	.5919 − 1	.3722
67°30′	1.178	.9656 − 1	.5828 − 1	.3828
68°0′	1.187	.9672 − 1	.5736 − 1	.3936
68°30′	1.196	.9687 − 1	.5641 − 1	.4046
69°0′	1.204	.9702 − 1	.5543 − 1	.4158
69°30′	1.213	.9716 − 1	.5443 − 1	.4273
70°0′	1.222	.9730 − 1	.5341 − 1	.4389
70°30′	1.230	.9744 − 1	.5236 − 1	.4509
71°0′	1.239	.9757 − 1	.5126 − 1	.4630
71°30′	1.248	.9770 − 1	.5015 − 1	.4755
72°0′	1.257	.9782 − 1	.4900 − 1	.4882
72°30′	1.265	.9794 − 1	.4781 − 1	.5013
73°0′	1.274	.9806 − 1	.4659 − 1	.5147
73°30′	1.283	.9817 − 1	.4533 − 1	.5284
74°0′	1.292	.9828 − 1	.4403 − 1	.5425
74°30′	1.300	.9839 − 1	.4269 − 1	.5570
75°0′	1.309	.9849 − 1	.4130 − 1	.5720
75°30′	1.318	.9859 − 1	.3986 − 1	.5873
76°0′	1.326	.9869 − 1	.3837 − 1	.6032
76°30′	1.335	.9878 − 1	.3682 − 1	.6197
77°0′	1.344	.9887 − 1	.3521 − 1	.6366
77°30′	1.353	.9896 − 1	.3353 − 1	.6542
78°0′	1.361	.9904 − 1	.3179 − 1	.6725
78°30′	1.370	.9912 − 1	.2997 − 1	.6915
79°0′	1.379	.9920 − 1	.2806 − 1	.7114
79°30′	1.388	.9927 − 1	.2606 − 1	.7320

Table 3   Logs of the Trigonometric Functions

Degrees	Radians	Log of Sin	Log of Cos	Log of Tan
80°0′	1.396	.9934 − 1	.2397 − 1	.7537
80°30′	1.405	.9940 − 1	.2176 − 1	.7764
81°0′	1.414	.9946 − 1	.1943 − 1	.8003
81°30′	1.422	.9952 − 1	.1697 − 1	.8255
82°0′	1.431	.9958 − 1	.1436 − 1	.8522
82°30′	1.440	.9963 − 1	.1157 − 1	.8806
83°0′	1.449	.9968 − 1	.0859 − 1	.9109
83°30′	1.457	.9972 − 1	.0539 − 1	.9433
84°0′	1.466	.9976 − 1	.0192 − 1	.9784
84°30′	1.475	.9980 − 1	.9816 − 2	1.0164
85°0′	1.484	.9983 − 1	.9403 − 2	1.0581
85°30′	1.492	.9987 − 1	.8946 − 2	1.1040
86°0′	1.501	.9989 − 1	.8436 − 2	1.1554
86°30′	1.510	.9992 − 1	.7857 − 2	1.2135
87°0′	1.518	.9994 − 1	.7188 − 2	1.2806
87°30′	1.527	.9996 − 1	.6397 − 2	1.3599
88°0′	1.536	.9997 − 1	.5428 − 2	1.4569
88°30′	1.545	.9999 − 1	.4179 − 2	1.5819
89°0′	1.553	.9999 − 1	.2419 − 2	1.7581
89°30′	1.562	.0000	.9408 − 3	2.0591
90°0′	1.571	.0000	undefined	undefined

Index

A

$\overrightarrow{AB}$, 184
$a > b$, 23
$a \geq b$, 23
$a + bi$, 220
$\sqrt[q]{a}$, 31
a^n, 29
a^x, 66
Absolute value, 225
Additive identity, 26
Adjacent side, 92
Ambiguous case, 166
amp(z), 234
Amplitude, 215
  of a complex number, 234
Angle, 103
  initial side of, 103
  in standard position, 104
  of depression, 99
  of elevation, 99
  terminal side of, 103
  vertex of, 103
antilog y, 75
Antilogarithm, 75
Argand diagram, 231
Associative laws for union and intersection, 14
Associativity, Axiom of, 27
Axiom of associativity, 27
Axiom of commutativity, 27
Axiom of distributivity, 28

B

$b < a$, 23
$b \leq a$, 23
Bar notation for repeating decimals, 21
Bar notation for logarithms, 75
Base, 29
Bearing, 160

C

C, 220
$C(x)$, 195
$c(x)$, 75
Characteristic of a logarithm, 75
Circular functions, 195
Coefficient, 27
Common logarithms, 75
Commutative law, 26
Commutativity, Axiom of, 27
Complement, 7
Complex conjugate, 225
Complex number(s), 220
  absolute value of, 225
  amplitude of, 234
  modulus of, 225
  nth root of, 239
  polar form of, 234
  trigonometric form of, 234
Constant function, 39
Constants, 26
Coordinate axes, 43
cos α, 93, 107